CELL SIGNALLING

CELL SIGNALLING

JOHN T. HANCOCK

DEPARTMENT OF BIOLOGICAL AND BIOMEDICAL SCIENCES
UNIVERSITY OF THE WEST OF ENGLAND

 LONGMAN

Addison Wesley Longman

Addison Wesley Longman Limited,
Edinburgh Gate, Harlow,
Essex CM20 2JE, England
and associated companies throughout the world

First published 1997

British Library Cataloguing in Publication Data
A catalogue entry for this title is available from the British Library

ISBN 0-582-31267-1

Library of Congress Cataloging-in-Publication Data
A catalog entry for this title is available from the Library of Congress

Set by 30 in 10pt Bembo
Printed in Great Britain by Henry Ling Ltd.,
at the Dorset Press, Dorchester, Dorset.

DEDICATION

To Sally-Ann and Thomas

CONTENTS

PREFACE

This book was written with the intention that it would be useful for degree students in their second and final years of undergraduate studies. It is also hoped that the book will be of interest to those on a medical degree, or those embarking on post-graduate studies, where they are perhaps new to this area or need to refresh earlier studies. The field of cell signalling is very large and continues to expand, with a great deal of new literature published in the journals every month. Here the aim is to describe the main components used by cells in their communications, with discussion on the ways in which such components might interact. This is not an approach taken by many books that cover this area of biochemistry, but it is becoming apparent that pathways are not just a single line of signal transducers leading simply from the original signal to the cellular response, but an intricate interplay of many components, with one pathway having profound effects on another. Therefore, the discussion of such components in isolation allows the knowledge gained about one particular component to be used when considering many pathways, not just the one in which it was first discovered. The aim is also to have a generic approach, as the ubiquity of some of these systems means that an understanding of the mechanisms is relevant to all organisms throughout nature and not just mammals or indeed the animal kingdom.

Throughout the book I have tried to give a reasonably comprehensive list of further reading which might be useful for the reader. However, it has proved impossible to do justice to every area in this field and the overall emphasis reflects my interests, so some people may feel that I have overlooked their particular area of research. Chapter 1 also contains references to the several excellent chapters published in major textbooks covering cell signalling, as well as other books published on this topic, and as such would be a good place to start additional reading.

The book starts with an overview chapter, encompassing many of the ideas and discussions applicable to all areas of cell signalling. Chapters 2 and 3 concentrate on the use of extracellular signals and their perception, while Chapters 4–8 focus on the intracellular components thought to be of most importance. Chapters 9 and 10 take more specific examples of signalling pathways and try to show how these components might come together to form complete pathways.

The content of the book was greatly shaped by many conversations with colleagues, in particular Steven Neill, who instigated a new course that inspired this book and to whom I express thanks, especially for reading the draft manuscript. I would also like to acknowledge help and support I received from Chris Pallister, Richard Luxton and Vyv Salisbury. Lastly, I would like to thank my wife Sally-Ann for her continued help and encouragement.

ABBREVIATIONS

AA	arachidonic acid
ABA	abscisic acid
ADP	adenosine diphosphate
AHL	N-acetyl-L-homoserine lactone
AMP	adenosine monophosphate
ARF	ADP-ribosylation factor
βARK	β-adrenergic receptor kinase
ATP	adenosine triphosphate
Boss	Bride of sevenless
cADPR	cyclic ADP ribose
cAMP	cyclic adenosine monophosphate (adenosine 3′,5′-cyclic monophosphate)
CAP	catabolite gene activator protein
cAPK	cAMP-dependent protein kinase (otherwise known as protein kinase A, PKA)
CAPP	ceramide-activated protein phosphatase
CCE	capacitative Ca^{2+} entry
cCMP	cyclic cytidine monophosphate (cytidine 3′,5′-cyclic monophosphate)
cDNA	complementary deoxyribonucleic acid
CGD	chronic granulomatous disease
cGI-PDE	cyclic GMP-inhibited phosphodiesterase
cGMP	cyclic guanosine monophosphate (guanosine 3′,5′-cyclic monophosphate)
cGPK	cGMP-dependent protein kinase
CIF	Ca^{2+} influx factor
cIMP	cyclic inosine monophosphate (inosine 3′, 5′-cyclic monophosphate)
CK	cytokinin
CRE	cAMP-response element
DAG	diacylglycerol
DGK	diacylglycerol kinase
DNA	deoxyribonucleic acid
EDRF	endothelium-derived relaxing factor
EDTA	ethylenediaminetetraacetic acid
EET	epoxyeicosatrienoic acid
EGF	epidermal growth factor
ER	endoplasmic reticulum
ERK	extracellular signal-regulated kinase
FAD	flavin adenine dinucleotide
FGF	fibroblast growth factor
FMN	flavin mononucleotide
FSH	follicle stimulating hormone
GA	gibberellin

GABA	γ-aminobutyric acid
GAP	GTPase-activating protein
G-CSF	granulocyte colony-stimulating factor
GDP	guanosine diphosphate
GEF	guanine nucleotide exchange factor
GM-CSF	granulocyte–macrophage colony-stimulating factor
GMP	guanosine monophosphate
GNRP	guanine nucleotide-releasing protein
GRK	G protein-coupled receptor kinase
GTP	guanosine triphosphate
HCR	haem-controlled repressor
HETE	hydroxyeicosatetraenoic acid
HPETE	hydroperoxyeicosatetraenoic acid
HRE	hormone response element
Hsp	heat-shock protein
IAA	indoleacetic acid
IFN	interferon
IGF-1	insulin-like growth factor 1
IL	interleukin
$InsP_3$	inositol 1,4,5-trisphosphate
$InsP_4$	inositol 1,3,4,5-tetrakisphosphate
$InsP_6$	inositol hexaphosphate
IRE	insulin-responsive element
IRS-1	insulin receptor substrate 1
ISPK	insulin-stimulated protein kinase
JA	jasmonic acid
JAK	janus kinase
JIP	jasmonate-induced protein
LH	luteinising hormone
LHC	light-harvesting chorophyll a/b complex
L-NAA	L-N^{ω}-aminoarginine
L-NMA	L-N^{ω}-methylarginine
LPA	lysophosphatidic acid
LPC	lysophosphatidylcholine
LPS	lipopolysaccharide
MAP	mitogen-activated protein
MAPK	mitogen-activated protein kinase
MAPKK	mitogen-activated protein kinase kinase
MEK	MAPK/ERK kinase
MEKK	MEK kinase
MLCK	myosin light chain kinase
mRNA	messenger ribonucleic acid
$NAADP^+$	nicotinate adenine dinucleotide phosphate
NAD^+	nicotinamide adenine dinucleotide (oxidised)
NADH	nicotinamide adenine dinucleotide (reduced)
$NADP^+$	nicotinamide adenine dinucleotide phosphate (oxidised)

NADPH	nicotinamide adenine dinucleotide phosphate (reduced)
NAP-1	neutrophil-activating protein 1 (otherwise known as IL–8)
NMR	nuclear magnetic resonance
NO	nitric oxide
NOS	nitric oxide synthase
PA	phosphatidic acid
PAP	phosphatidate phosphohydrolase
PC	phosphatidylcholine
PCR	polymerase chain reaction
PDE	cyclic nucleotide phosphodiesterase
PDGF	platelet-derived growth factor
PE	phosphatidylethanolamine
PGHS	prostaglandin G/H synthase or cyclooxygenase
PH	pleckstrin homology
P_i	inorganic phosphate
PKC	protein kinase C
PLA_2	phospholipase A_2
PLC	phospholipase C
PLD	phospholipase D
PMA	phorbol 12-myristate 13-acetate
PMCA	plasma membrane Ca^{2+}-ATPase
PP_i	inorganic pyrophosphate
PtdIns	phosphatidylinositol
PtdIns 4-P	phosphatidylinositol 4-phosphate
PtdIns P_2	phosphatidylinositol 4,5-bisphosphate
PTP	protein tyrosine phosphatase
RTK	receptor tyrosine kinase
RyR	ryanodine receptor
SDS-PAGE	sodium dodecyl sulphate–polyacrylamide gel electrophoresis
SERCA	smooth endoplasmic reticulum Ca^{2+}-ATPase
Sev	Sevenless
SH2 domain	Src homology domain 2
SH3 domain	Src homology domain 3
SM	sphingomyelin
Sos	Son of sevenless
STAT	signal transducers and activators of transcription
TGF	transforming growth factor
TNF	tumour necrosis factor
TPA	12-O-tetradecanoyl-phorbol-12-acetate (otherwise known as PMA)
VSP	vegetative storage protein

1 Aspects of Cellular Signalling

Topics

- What Makes a Good Signal?
- Different Ways Cells Signal to Each Other
- Coordination of Signalling
- Domains and Modules
- Oncogenes
- A Brief History
- Techniques in the Study of Cell-Signalling Components

Introduction

The ability of organisms, or individual cells within an organism, to sense and respond to their environment is crucial to their survival. All cells must have the ability to detect the presence of extracellular molecules and conditions, and must also be able to instigate a range of intracellular responses. Such systems have to be carefully orchestrated and controlled. To enable living organisms to do this, a complex and interwoven range of signalling pathways has evolved. The signalling systems of a single-cell organism are complex, but a multicellular organism has to coordinate the functioning of cells that may be a huge distance apart, possibly in the order of metres. Such cells may also have different and extremely specific functions.

The main components and principles underlying the signalling mechanisms are basically the same across the diverse range of organisms (bacteria, plants and animals) with individual components providing the specificities or functions required for a particular cell. Knowledge of a signalling system in one species is often used to speed the discovery of an analogous system in a completely different species, something that only a few years ago would have been thought of as impossible.

Cell signalling is not only important for understanding the function of a normal cell, but is also of vital importance in understanding the growth and activity of an aberrant cell. The discovery of oncogenes, genes causing the uncontrolled growth of cells that may lead to cancerous growths, was heralded as a major breakthrough in the understanding of cancer in humans, while the discovery of cytokines held great hopes as a cure for a variety of diseases.

Once the function of a protein is elucidated, researchers often turn to the control of its activity and the control of its synthesis to get a better understanding of how the protein fits into the overall biochemistry of the cell. The literature is thus expanding at an alarming rate owing to the new signalling mechanisms coming to light, with several journals dedicated to the publication of research in this field.

The aim of this book is to discuss the main components found in cell-signalling pathways, rather than describe the many specific pathways that might contain common elements. The details in the text are not directly referenced, but further reading is given at the end of each chapter to enable the reader to refer to the original research or to review articles on specific areas.

What Makes a Good Signal?

Signals seem to come in various shapes but they generally have a set of common characteristics. To be an effective signal, a molecule has to be relatively small and able to travel from the site of manufacture to the target site. Extracellular signals can take advantage of the vascular system of the organism to assist in their travels. Inside the cell the signal molecule often appears to rely on diffusion, although the cellular cytoskeleton is thought to play a part. Therefore, intracellular signals need to be small enough to diffuse rapidly and frequently need to have the capacity to diffuse through membranes, either with the aid of a carrier protein or using their hydrophobic nature.

A second characteristic that signals possess is that they can be made or altered relatively quickly. For example, cyclic adenosine monophosphate (cAMP) is produced enzymatically inside the cell while calcium ions can be released from intracellular stores. Often the cell has to have a fast response and therefore the signal has to be relayed through the cell as quickly as possible. However, once it has been produced and had its effect, a signal must have the capacity to be turned off again. Frequently, the cessation of a signal also needs to be rapid. If a cell is instructed to produce a certain gene product having received an extracellular message, it will not be required to produce that product forever until the cell dies. The stimulation of gene expression needs to be reversed to allow the cell to return to its unstimulated state. Although cells have developed an array of enzymes involved in turning off signals, such as phosphodiesterases and phosphatases, generally it is this part of the pathway that is poorly understood. For example, many studies show that a kinase is required to stimulate a certain cellular activity, but often the reversal of the stimulation is brushed aside or the molecular mechanism is assumed. In fact, it is almost certainly the balance between the on and off signals that is important.

Different Ways Cells Signal to Each Other

Cells can signal to each other in many different ways, mainly dependent on the distance between the signalling cell and the target cell. If the cells are touching, signalling may simply be through pores in the membranes, such as gap junctions or plasmodesmata, or due to a membrane-bound ligand being identified by a receptor in the membrane of a neighbouring cell. If the cells are further apart, they may communicate via the release of molecules that are then detected by the target cell or via the transmission of an electrical signal (Figure 1.1).

Figure 1.1
Three of the ways in which cells communicate with each other: (a) the release of molecules by one cell that are detected by a receptor on another cell; (b) the detection of a membrane protein on the surface of one cell by a receptor on a second cell; (c) the direct transfer of small molecules through a gap junction

(a)

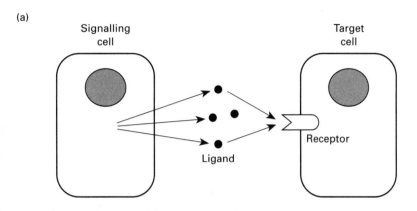

(b)

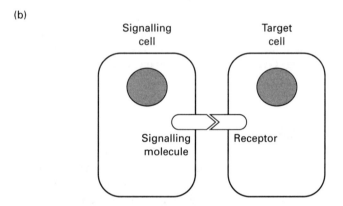

(c)

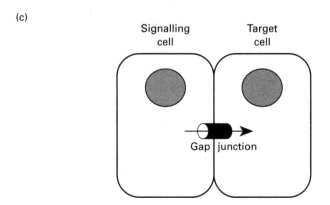

The term 'ligand' is used throughout this book as a generic term for any molecule that binds to specific sites on a protein, such as a hormone binding to its receptor. The word 'ligand' comes from Latin *ligare*, which means to bind.

Synaptic

A fast and efficient method of signalling over long distances is via changes in the electrical potential across the membrane of the cell, such as in the nervous system in animals. Such a potential is propagated along the length of the cell and can therefore travel considerable distances in a body, as the cells are extraordinarily long. These cells, known as neurones, usually contain four regions: the axon, for conduction of the signal; a cell body, containing the nucleus and normal cellular functions; dendrites, which receive chemical signals from other neurones or other cells; and axon termini, from where the signal is passed on to the next neurone or target tissue. Neurones communicate through the use of synapses. These are areas of close contact through which the signal can be propagated. Synapses generally fall into two groups, transmitting the electrical signal either directly through gap junctions or chemically by the release of neurotransmitters that are detected by the target cell.

Although electrical signalling was always thought to be unique to animals, plants also use a primitive type of electrical impulse in response, for example, to wounding. Over relatively long distances the signals are carried by the vascular system, but short signalling distances involving the plasmodesmata have also been seen.

Endocrine

In endocrine signalling, cells release signalling molecules, such as hormones, that can travel vast distances in the organism, usually having an effect in a different tissue. In animals, the bloodstream supplies the route to carry these messengers, but it is relatively slow and unspecific. All the tissues will be bathed in a supply of the signalling molecule and specificity comes from the target cell's ability to detect it. Plants also use endocrine signalling, hormones once again being carried by the vascular system of the organism. However, even unicellular organisms have been shown to rely on hormonal signalling, amoebae relying on diffusion in the surrounding medium to carry the signal to the next organism.

Paracrine

As seen with endocrine signalling, a common way for cells to communicate is by the release of molecules that diffuse to, and are detected by, a second cell. In paracrine signalling the released molecules are detected and have their effect in the local area of the signalling cell. Diffusion of the signalling molecules is very limited because they are either rapidly destroyed by extracellular enzymes or rapidly immobilised on, or taken up by, neighbouring cells. An example of such signalling is the release of neurotransmitters from neurones, which have their effect in the neighbouring neurone or in the neighbouring cells such as a muscle cell.

Autocrine

In a similar way to paracrine, autocrine describes a local effect of diffusable signalling molecules, but here the term 'autocrine' is specifically used to mean released signalling molecules that act on the same cell that released them. This type of signalling is common in growth hormones and sometimes with molecules in the eicosanoid group. It is often found in cells that are in a developmental or differentiating stage and can be seen as an emphasising feature: once a cell has been instructed to follow a certain developmental route, an autocrine signal may ensure that the set route is followed by releasing a signal to emphasise that message. Autocrine signalling is also commonly found in tumour cells which produce growth hormones that stimulate the uncontrolled growth and proliferation.

Direct Cell–Cell Signalling

Cells that are in physical contact with each other are often very active in signalling to each other. They can achieve this either by the recognition of molecules on each others' surfaces or by direct communication through specialised areas of the cell surface.

Receptor–ligand signalling

A common way that cells communicate is through the recognition of surface markers (see Figure 1.1b). A good illustration of this is the communication of cells in the eyes of the fly *Drosophila* during their development. The fly's compound eye contains approximately 800 separate photoresponsive units, or ommatidia, which each consist of 22 cells. Eight of these are responsible for photoreception and are called retinula (R) cells. In the development of the ommatidia, the R8 cells express on their surface a protein known as Bride of sevenless (Boss). The presence of this protein is detected by a receptor tyrosine kinase known as Sevenless (Sev) on the surface of the neighbouring R7 cell. The binding of Sev to the Boss protein triggers the R7 cells to develop and enables the fly to detect ultraviolet light. This system was discovered due to mutations in the Sev protein where no R7 cells were signalled to develop, hence the naming of the protein as Sevenless. Such flies were identified by their inability to detect ultraviolet light and such animals have been extremely useful in the elucidation of signalling pathways.

Other proteins involved in this type of cell–cell interaction identified in *Drosophila* include Notch, Armadillo and the product of the *dlg* gene; such proteins may help to control cell proliferation.

5

Gap junctions and plasmodesmata

In animal tissues, cells are often seen to have an area where the two plasma membranes of adjoining cells are apparently held at a fixed but very small distance apart. These areas are known as gap junctions. The plasma membranes in this area contain proteins that form tubes; when the tubes in the two membranes come into perfect alignment a complete pipe is formed, which allows the passage of small molecules directly from one cytoplasm to the next. The passage of molecules of less than 1200 Da, for example signalling molecules such as cAMP or Ca^{2+} ions, can therefore take place directly from one cell to the next (see Figure 1.1c).

The tubular structures in the membranes, known as connexons, are composed of a group of six identical proteins called connexins arranged in a ring. The connexons protrude slightly from the membrane and hence the two membranes are held apart by the same distance (i.e. the distance of the protrusion from each cell added together) (Figure 1.2). Each gap junction may contain several hundreds of connexons, which can be revealed by freeze-fracture techniques.

Figure 1.2
Schematic representation of a gap junction showing the alignment of the connexin proteins forming connexons.

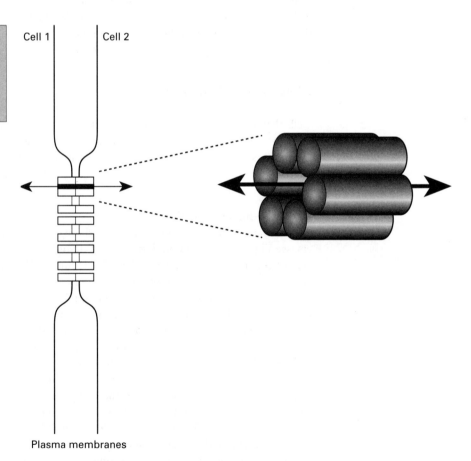

Cell 1   Cell 2

Plasma membranes

The protein structure of connexins has been predicted from the amino acid sequences. At least 11 isoforms coded for by separate genes have been identified and therefore conserved features have been used to determine the likely topology of the polypeptides. It has been predicted that each protein contains four membrane-spanning helices, with a particularly conserved helix sequence being used to line the hole that is made when the six subunits come together.

The expression of the isoforms differs between tissues as does the permeability of gap junctions. However, a single cell can contain more than one isoform but the 12 connexins of a single junction are of the same type.

Passage through gap junctions is not a free-for-all but very well controlled. These junctions are rapidly closed if the Ca^{2+} or H^+ concentration rises. Connexins undergo a reversible conformational change, which probably works in a similar manner to the aperture of a lens. Closure of gap junctions is of particular importance if the neighbouring cell dies: separation from the cytosol of a cell which has limited or no control of the input of molecules would be vital to the other cell's survival.

Plant cells also have direct cell–cell connections but owing to the presence of a cell wall they differ from gap junctions. The connections in plants are called plasmodesmata (Figure 1.3). The cell walls contain holes that are lined by the plasma membrane of the cells and in fact the membrane can be perceived as continuous. This lined hole is approximately 20–40 nm in diameter and allows the cytosol to be contiguous with the neighbouring cell. This intimacy is continued, as passing through the hole is a membranous tube called the desmotubule, which is derived from the endoplasmic reticula of the cells. Therefore a cell's cytosol, plasma membrane and endoplasmic reticulum are shared with the neighbouring cell.

Figure 1.3
Schematic representation of a plasmodesma showing how the endoplasmic reticulum and cytosol are continuous across the two cells.

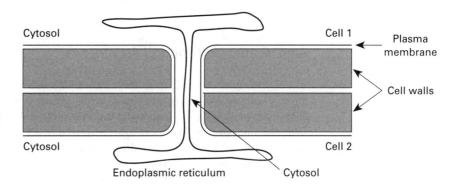

Once again this does not mean the free passage of all molecules between the cells. The cut-off is thought to be approximately 800 Da, allowing the passage of small signalling molecules but not macromolecules such as proteins, although interestingly the transfer of some proteins and even viruses has recently been reported. Passage through plasmodesmata is also under tight regulation but the processes controlling this are poorly understood.

Co-ordination of Signalling

One of the major puzzles in the field of cell signalling is how a cell coordinates all the signals it has to deal with. Researchers talk about integration of signalling pathways and 'cross-talk', but little concrete evidence is known about how the cell manages different signal pathways simultaneously. It is quite possible, for example, for a cell to detect two or more signals on the outside via its receptors. These receptors will then activate the relevant signalling pathways inside the cell, but quite often these signalling pathways will have common elements. For example, two signals may both use modulation of the intracellular Ca^{2+} ion concentration as one of their means of signalling. How does the cell determine if the alteration of Ca^{2+} is due to the activation of receptor 1 or receptor 2? Or, indeed, is interaction of the signalling pathways the key? Does the cell rely on the subtle changes made to one signal by a second signalling pathway and does this give the cell the fine control required? Often, the end-result of the activation of receptors may in fact be different or even opposite, and yet both pathways when studied in isolation may appear to have common components. Another example of such a dilemma is seen when studying the signalling induced in insect gut muscles, where two receptors that apparently have differing effects on the muscle cells both induce the production of diacylglycerol (DAG) and presumably the activation of protein kinase C (PKC). How does the cell know what the production of DAG is supposed to mean? Researchers discuss the existence of compartmentalisation of cells and the role of structural scaffolds inside cells, which restrict the movement of signalling components through the cytoplasm (discussed below). It may need an alteration in our perception of what a cell actually looks like on the inside before the answers to these problems are solved, but it is surely a debate that will continue.

Amplification and Physical Architecture

The mechanisms of signal transduction from one cell to the next and subsequently into the interior of the target cell involve the formation of chains of signalling molecules, each passing on the message to the next molecule in the line. An extracellular signalling molecule, or first messenger, perceived by a cell often leads to the production of small and transient signalling molecules on the inside of the cell, called second messengers. Such second messengers would then activate or alter the activity of the next component of the transduction pathway, for example a kinase. A good example of a second messenger is cAMP, as discussed in Chapter 5.

One of the important features of the production of second messengers and the existence of these cell-signalling cascades is the capacity to allow amplification of the original signal, so that the end-result is effective in the target cell. Binding of a single hormone molecule to a receptor on the cell surface does not generally cause activation of a single enzyme molecule. Rather, a situation could be envisaged as shown in Figure 1.4.

Here, binding of ligand causes the receptor to switch on many enzyme molecules on the plasma membrane. Such transduction may involve G proteins acting on enzymes such as adenylyl cyclase. The activation of one adenylyl cyclase polypeptide leads to the production of many cAMP molecules causing further amplification of

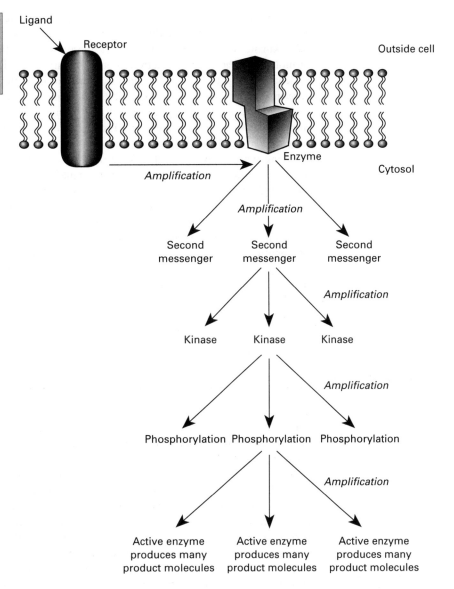

Figure 1.4
An example of how amplification could take place in a signal-transduction cascade.

Ligand

Receptor

Outside cell

Enzyme

Cytosol

Amplification

Amplification

Second messenger | Second messenger | Second messenger

Amplification

Kinase | Kinase | Kinase

Amplification

Phosphorylation Phosphorylation Phosphorylation

Amplification

Active enzyme produces many product molecules | Active enzyme produces many product molecules | Active enzyme produces many product molecules

the signal. These cAMP molecules in turn activate many kinases, which phosphorylate a great many proteins, causing yet further amplification of the signal. Therefore, it can be seen that the binding of one hormone molecule to the exterior surface of the cell can potentially lead to the activation of hundreds or thousands of enzyme molecules inside the cell, each of which catalyse many rounds of a reaction.

However, as discussed above, signal cascades are not straightforward pathways with a defined and invariable route. Most signalling molecules can have an effect on several different signalling systems in the cell, so that the stimulation of a single element of the pathways may lead to the further stimulation of several other signalling branches. Therefore, it can be seen that signals have the potential for both amplification and divergence. For example, a kinase may lead to the phosphorylation of several proteins that have very different effects in the cell, including the activation of other kinases. Likewise, the activation of phospholipase C (PLC) leads to the release of molecules that ultimately activate both PKC and Ca^{2+} signalling pathways. Conversely, several pathways may lead to the stimulation of the same enzyme, showing convergence of pathways. An example of this would be the breakdown of glycogen, which can be stimulated by the arrival of a hormone at the cell surface, leading to intracellular signalling through cAMP or, separately, by the release of intracellular stores of Ca^{2+}.

Along with the proteins of defined signalling function, other polypeptides have also been postulated to be involved in the signalling systems of cells. Such proteins are thought to have a scaffolding role by holding the relevant components of a cascade in a particular physical environment, allowing efficient transfer of the signal but in doing so restricting their interaction with other structurally and functionally similar components in a related signalling cascade. One such protein is the STE5 polypeptide of yeast. However, such a restriction on movement of components would place a question mark over the potential amplification or divergence of the signal that could take place. It is possible that cells have evolved a pay-off, whereby amplification is compromised in exchange for the need to retain vital specificity of a signal-transduction cascade.

Domains and Modules

The details of signalling pathways of cells look incredibly complex but quite often components contain areas of similarity or homology that are repeated. Different kinases, for example, might have sequences that can be recognised as the catalytically active regions of the polypeptide and, although not the same in different proteins, a consensus for the sequence in that area can be postulated. Other areas may be attached on to this basic structure, such as sequences responsible for regulating the activity. Again, this second area may share homology with regions in other proteins controlled in the same or similar ways but which contain a different catalytic region and therefore have a completely different function. It can be seen that these polypeptide sequences and regions are like functional modules that can be glued together in an incredibly varied arrangement. Of course, this is a gross simplification of the situation, but a good illustration is seen with the two-component signalling system involving the histidine kinase activity of prokaryotes, as described in Chapter 4.

Two very important polypeptide domains found in many proteins that have a role in cell signalling are the Src homology domains, the so-called SH2 and SH3 domains. Both of these are involved in the interactions of proteins, acting as a glue to hold them together; however, the specificity of the two types of sequence differs.

SH2 domains are in general approximately 100 amino acids in length and have a binding affinity for phosphotyrosine amino acids, i.e. tyrosine residues of a protein that have been phosphorylated. The domain has the structure of a deep pocket lined with positively charged amino acids, which interact with the phosphate group of the phosphotyrosine. Binding is very much reduced if dephosphorylation has occurred.

A second region of importance of SH2 domains has been used to categorise them into two groups. The first group also contains a second binding pocket that is partly responsible for bestowing specificity on the binding, as it recognises a single amino acid of the target polypeptide. The second SH2 group contains a region that is indented and shows binding to a group of mainly hydrophobic amino acids, a little way removed from the target phosphotyrosine residue.

SH3 domains, on the other hand, bind to a sequence on the target polypeptide containing several proline residues, which often have a left-handed helical secondary structure known as a polyproline helix type II.

It is quite common for a single protein to contain both of these domains and even multiple copies of them. They are particularly common where proteins that are not normally associated have to come together, leading to activation of one of the proteins. An excellent example of this is provided by adaptor proteins, such as GRB2. GRB2 is a 25 kDa polypeptide that contains one SH2 domain and two SH3 domains and which can be associated with guanine nucleotide-releasing proteins (GNRPs), for example Son of sevenless (Sos). These proteins are often involved in signalling from receptor tyrosine kinases, such as the epidermal growth factor (EGF) receptor, which are autophosphorylated on binding to their ligand. The formation of phosphotyrosine residues on the receptor then allows the binding of these adaptor proteins through their SH2 domains, and in so doing causes the activation of their associated guanine nucleotide-releasing proteins, resulting in the stimulation of a G protein and the activation of the rest of the cascade. Other adaptor proteins include Shc, Crk and Nck, although to date these are not as well characterised as GRB2.

Another example of the involvement of SH2 domains is on the STAT proteins (**s**ignal **t**ransducers and **a**ctivators of **t**ranscription). These are phosphorylated by Janus kinases (JAKs), which become associated with activated cytokine receptors. The STAT proteins are phosphorylated on tyrosine and because they also contain SH2 domains they dimerise. The dimer so formed then proceeds to cause the enhancement of transcription. At least three STAT proteins have been identified: STAT1α (91 kDa) and STAT1β (84 kDa) arise from the same gene via alternative splicing while STAT2 is slightly larger at 113 kDa.

Other proteins may not contain such domains themselves but on becoming phosphorylated gain the topology to interact with such domains resident on other proteins. These proteins, often referred to as relay proteins, are also seen in signalling cascades from receptor tyrosine kinases. A good example is insulin receptor substrate 1 (IRS-1). This protein is phosphorylated on multiple tyrosines, so creating sites that can interact with the adaptor protein GRB2 via its SH2 domain. This particular pathway is discussed in more detail in Chapter 9.

Other similar protein-binding domains, such as the pleckstrin homology (PH) domain, have also been recognised recently. PH domains appear to be a rather diverse group of structural motifs and no specific binding site for PH domains has been identified. The true picture may be that there are several separate subtypes of PH domains, each with their own specificity. Structural analysis of some PH domains suggests that the polypeptide is folded into a barrel of β-sheet with a single α-helix, but many variations of this basic structure have been seen.

It is probable that more binding domains will be found as further cloning and sequencing work allows consensus sequences to be deduced.

Oncogenes

The discovery of oncogenes and their functions was heralded as a great advance in our fight against cancer and in the understanding of why tumour cells form. The first hint of what was going on was supplied as far back as 1911 by Peyton Rous, with his work on chicken tumours and his discovery of tumour viruses. Even today, many of the oncogenes that have been characterised were originally isolated from retroviruses. It is now clear that most oncogenes code for proteins which have an influence on cellular signalling mechanisms, affecting the way a cell perceives an extracellular signal or indeed the way in which that signal is transmitted through the cell. As such, oncogenes can be grouped into four classes. Class 1 includes oncogenes coding for growth factors, such as the *sis* gene that codes for platelet-derived growth factor (PDGF). Oncogenes encoding growth factor receptors make up class 2 and include receptors for EGF, the *erbB* gene, and nerve growth factor, the *trk* gene. Oncogenes coding for intracellular signalling components make up the third and fourth classes, the class into which they fit being determined by their intracellular location. Class 3 includes the G protein gene *ras* as well as the kinase genes *src* and *raf*; all are cytosolic proteins. The nuclear proteins that make up class 4 include transcription factors, such as *jun* and *fos*, and steroid receptors such as the *erbA* product.

Oncogenes may actually encode defective components of a signalling pathway, which in many instances are in a permanently activated state. For example, the *erbB* gene codes for a receptor that has been truncated at both ends and signals to the cell the presence of EGF even in its absence. Likewise, the *ras* oncogene traps the G protein in a guanosine triphosphate (GTP)-bound state, where GTP hydrolysis cannot take place and the G protein remains permanently active.

Although many oncogenes are virally coded, many are coded for by our own genomes. These genes, known as protooncogenes, are normally quiescent but may be turned on at a later date. The resultant overproduction of, for example, a growth factor will then stimulate the overgrowth of the cells and lead to the formation of tumours.

A Brief History

A quick glance through a current molecular biology journal will reveal the magnitude of interest in cell-communication systems and signal-transduction pathways, with the literature expanding at what seems like an exponential rate. However, this interest is not new and has spanned the period of biochemical research. Even in the middle of the last century, Darwin was experimenting with the phototropism of coleoptiles and suggested the presence of a substance that was transported around plants, having its effect at a site distant from its site of production and secretion. This substance was discovered by Went in 1928 and was later identified as indoleacetic acid or, to give it its common name, auxin.

In an animal context, the term 'hormone' (from the Greek meaning to excite or arouse) was introduced by Starling as long ago as 1905. However, lipid-soluble hormones such as prostaglandins had to wait 30 years to be discovered. This family of compounds was originally named after the organ supposed to be their site of production but this has subsequently proved to be misleading.

Although extracellular events were beginning to be unravelled, some of the first clues as to what was happening inside cells came in the late 1940s and 1950s. In 1947, experiments in which small quantities of Ca^{2+} ions were injected directly into the cell showed that an increase in intracellular Ca^{2+} led to skeletal contractions. In the 1950s Earl W. Sutherland and Theodore Rall, working at Western Reserve University, found that the addition of epinephrine and glucagon to cells led to the binding of these hormones to cell receptors, and experimentation showed the subsequent production of cAMP. The enzyme responsible for this increase in intracellular cAMP concentration they named adenyl cyclase, but it is now known as adenylyl cyclase or adenylate cyclase. However, it was not until 1989 that Krupinski and colleagues cloned the first mammalian enzyme. Adenosine monophosphate (AMP) and guanosine monophosphate (GMP) were isolated from rat urine in 1963 while guanylyl cyclase was identified in the late 1960s. Investigations of the role of adenylyl cyclase and cAMP was further helped by the isolation of adenylyl cyclase-deficient cells, known as cyc⁻. These cells were first reported in 1975 by Gordon Tomkins, Henry Bourne and Philip Coffinio working at the University of California.

The involvement of a separate signalling pathway, that of the inositol phosphate pathway, was also starting to unravel in the 1950s. The role of phospholipids, such as phosphoinositides, was first suggested in 1953 when Hokin and Hokin showed that the addition of acetylcholine to cells was found to lead to the incorporation of ^{32}P into what was regarded as a minor lipid, phosphatidylinositol. This pathway was shown to involve muscarinic receptors. The pathway leading to the biosynthesis of these lipids in the endoplasmic reticulum was reported by Agranoff and Paulus and their respective colleagues in the late 1950s. However, a link was made to another signalling pathway in 1975, when Michell suggested that messengers which stimulated the breakdown of phosphatidylinositol also caused an increase in cytosolic Ca^{2+}. A new pathway involving membrane inositol lipids was more recently suggested when in 1988 Whitman first identified phosphatidylinositol 3-phosphate in transformed lymphocytes. This is now an exciting new area of lipid signalling that is implicated in the signal transduction of several systems, including responses to insulin. Similarly, evidence in 1986 that sphingosine inhibited PKC has led to elucidation of the sphingomyelin cycle.

Protein phosphorylation is the common end-result of many signal pathways, and it was over 30 years ago that regulation of glycogen breakdown by phosphorylation of glycogen phosphorylase was suggested. In this system, the protein kinase was itself regulated by the intracellular cAMP concentration, but 1970 saw the first reports of a kinase more sensitive to cyclic guanosine monophosphate (cGMP) than cAMP, i.e. cGMP-dependent protein kinase (cGPK). Until 1980, it was thought that only serine and threonine were targets for phosphorylation of proteins, but it is now clear that one of the major phosphorylation events is the addition of the phosphoryl group to tyrosine residues. However, it was not until 1988 that Tonks obtained the first partial sequence of a tyrosine phosphatase. Today, it has been suggested that the human genome might contain as many as 2000 kinase genes and up to 1000 phosphatase genes.

The history of the elucidation of the G protein story takes us back to the 1970s. Gilman, working at the University of Virginia, identified G proteins as crucial intermediates in signal transduction in the latter part of the decade and in 1980 Paul Sternweis and John Northup, working in Gilman's laboratory, purified the G protein G_S. It was not long afterwards that Lubert Stryer and Mark Bitensky independently discovered the existence of a second G protein, this time using rod cells from the eye; this G protein was transducin or G_t.

Today, it would be folly to claim that all the signalling pathways have been discovered, even if we are not sure of how many of them function. For example, it was not until 1987 that Moncada suggested that endothelium-derived relaxing factor was in fact nitric oxide, opening up a whole new field of research. However, with the advances in molecular genetic technology, the rapid rate of cloning of new genes and the Human Genome Project, new members of existing pathways will eventually come to light but the challenge to understand how all these pathways interact will remain.

Techniques in the Study of Cell-signalling Components

Understanding of the molecular events of cell signalling, like all areas of biochemistry, has been facilitated by the techniques available at the time. Early studies were based on observable events with whole cells, whereas today molecular biology allows us to discover new prospective signalling molecules without even knowing their function.

Immunological techniques have been of great benefit in the discovery and study of the distribution of receptors; coupled with modern fluorescence techniques, the receptors on a cellular surface may be visualised as a three-dimensional image using confocal technology. Photoreactive radioactive analogues of ligands have also been used in the study of, for example, lysophosphatidic acid (LPA) receptors. Antibodies raised against particular epitopes on a protein have been invaluable in the elucidation of the interactions between polypeptides. Also, by knocking out a particular active site, antibodies can stop a signal-transduction pathway and the effect on the cell or cellular function, such as transcription, can be observed. Similarly, if particular polypeptides are thought to undergo an interaction, short peptides can be synthesised in order to flood a prospective docking site on one polypeptide and thus potentially upset the propagation of the signal or activity.

One of the most commonly used techniques in the study of signalling pathways is the addition of stimulators or inhibitors. The question of what happens inside the cell on addition of a particular receptor ligand is still commonly asked. Once a cellular event has been demonstrated, the method by which it can be inhibited is soon addressed. Many of the inhibitors used, when discovered, were thought to be quite specific, but subsequent studies often show that their specificity can be doubted and that they have effects on more than one pathway. Interpretation of the results then becomes more difficult, and the search for better inhibitors continues.

Once a component of a pathway has been identified, the search for an understanding of its exact mechanism usually requires its purification. If the full primary structure of the polypeptide is known, the secondary and tertiary structures can either be predicted by computer analysis or the pure protein can be analysed by the time-consuming methods of nuclear magnetic resonance (NMR) or X-ray diffraction. Once several components of a system have been isolated, they can be recombined in a cell-free reconstitution system where the concentrations and conditions can be carefully controlled and altered.

However, the greatest advances made in the discovery of new signalling components and the study of their interactions have come through the use of molecular genetic techniques. Antisense technology can be used to specifically knock out the expression of elements in a pathway, while an antibody or the back-translation of an N-terminal protein sequence into an oligonucleotide can be used to extract a clone from a DNA library. This prevents the need for much of the protein sequencing undertaken previously. Oligonucleotide sequences based on the consensus sequences of several related polypeptides or complementary DNA (cDNA) clones can be used to find the genes for more members of a polypeptide family, where the functions have not even been discovered. Such DNA sequences can be ligated into plasmids and overexpressed in cells, so enhancing or reducing the pathway in the cell. For example, the overexpression of a defective kinase in a cell means that the defective enzyme preferentially receives the message from the signal pathway but is unable to undergo any kinase activity, so effectively stopping the pathway at that point. Similarly, the overexpression of a specific phosphatase means that a specific kinase activity may be negated.

Today, the logic of computer networks is being used to try to understand the complexity of the interaction of signalling pathways, where it is thought that combinations of certain pathways and their relative contributions to the overall signal may be important.

Summary

The potential for an organism or individual cell to signal to its neighbouring organism or cell and to detect and respond to such signals is crucial for its survival. Signals used in biological systems appear to be very diverse, ranging from a simple change in the concentration of an intracellular ion such as Ca^{2+} to complicated compounds; however, the production of the signal usually needs to be reasonably rapid, particularly for intracellular signals, and it must be able to be conveyed from its site of production to its site of action. Also, importantly, it must be readily reversible.

Signals between cells can be via electrical potential changes (synaptic signalling), the release of compounds detected over relatively large distances or even by the cell producing the signal (paracrine, endocrine or autocrine) or the passage of small chemicals through pores in the cell (gap junctions and plasmodesmata).

Signalling pathways can diverge to cause a multitude of cellular changes in response to a single signal; alternatively they can converge, where two or more signals effectively control the same metabolic pathway for example. However, in nearly all signalling pathways, a great deal of amplification takes place, allowing a small number of signalling molecules to precipitate a large effect.

Often the molecular study of protein components has revealed consensus patterns within polypeptides and many of the proteins can be considered to be made of functional modules. Many of these components have been identified as products of oncogenes, highlighting the importance of these signalling components in the control of cellular proliferation and differentiation, and the study of the functioning of these gene products has helped enormously in the elucidation of many pathways.

Although the study of cell-signalling cascades has been taking place for many years, the development of modern molecular biological techniques has allowed the further understanding of how many of these proteins function and even predicts the existence of isoforms of protein yet to be discovered.

Further Reading

Alberts, B., Bray, D., Lewis, J., Raff, M., Roberts, K. and Watson, J. D. (1994) *Molecular Biology of the Cell* (3rd edn). Garland Publishing, New York. Chapter 15 in particular.

Barritt, G. J. (1994) *Communications Within Animal Cells*. Oxford Science Publications, Oxford.

Bowles, D. J. (ed.) (1994) *Molecular Botany: Signals and the Environment*. Portland Press, London.

Hardie, D. G. (1991) *Biochemical Messengers: Hormones, Neurotransmitters and Growth Factors*. Chapman & Hall, London.

Heldin, C.-H. and Purton, M. (eds) (1996) *Signal Transduction*. Chapman & Hall, London. A collection of 23 chapters by experts in the field.

Lodish, H., Baltimore, D., Berk, A., Zipursky, S. L., Matsudaira, P. and Darnell, J. (1995) *Molecular Cell Biology* (3rd edn). W. H. Freeman, New York. Chapter 20 in particular.

Milligan, G. (1992) *Signal Transduction: A Practical Approach*. IRL Press, Oxford.

Morgan, N. G. (1989) *Cell Signalling*. Open University Press, Milton Keynes.

Stryer, L. (1995) *Biochemistry* (4th edn). W. H. Freeman, New York. An excellent general biochemistry text.

Different Ways Cells Signal to Each Other

Le Roith, D., Shiloach, J., Roth, J. and Lesniak, M. A. (1980) Evolutionary origins of vertebrate hormones: substances similar to mammalian insulins are native to unicellular eukaryotes. *Proceedings of the National Academy of Sciences* USA, **77**, 6184–8.

Parkinson, J. S. (1993) Signal transduction schemes of bacteria. *Cell*, **73**, 857–71.

Thain, J. F. and Wildon, D. C. (1992) Electrical signalling in plants. *Science Progress*, **76**, 553–64.

Woods, D. F. and Bryant, P. J. (1993) Apical junctions and cell signalling in epithelia. *Journal of Cell Science*, **17**, 171–81.

Co-ordination of Signalling

Bygrave, F. L., Karjalainen, A. and Hamada, Y. (1994) Crosstalk between calcium-mediated and cyclic AMP-mediated signalling systems and the short term modulation of bile flow in normal and cholestatic rat liver. *Cellular Signalling*, **6**, 1–9.

Houslay, M. D. (1995) Compartmentalisation of cyclic AMP phosphodiestereases, signalling crosstalk, desensitisation and the phosphorylation of G(I)-2 add cell specific personalisation to the control of the levels of the second messenger cyclic AMP. *Advances in Enzyme Regulation*, **35**, 303–38.

Amplification and physical architecture

Herskowitz, I. (1995) MAP kinase pathways in yeast: for mating and more. *Cell*, **80**, 187–97. Discussion of scaffolding proteins.

Domains and Modules

Alex, L. A. and Simon, M. L. (1994) Protein histidine kinases and signal transduction in prokaryotes and eukaryotes. *Trends in Genetics*, **10**, 133-8. Illustration of protein modules.

Cohen, G. B., Ren, R. and Baltimore, D. (1995) Modular binding domains in signal transduction proteins. *Cell*, **80**, 237–48.

Feller, S. M., Ren, R., Hanufusa, H. and Baltimore, D. (1994) SH2 and SH3 domains as molecular adhesives: the interactions of Crk and Abl. *Trends in Biochemical Science*, **19**, 453–8.

Koch, C. A., Anderson, D., Ellis, C., Moran, M. F. and Pawson, T. (1991) SH2 and SH3 domains: elements that control interactions of cytoplasmic signalling proteins. *Science*, **252**, 668–74.

Matsuda, M., Mayer, B. J. and Hanafusa, H. (1991) Identification of domains of the v-*crk* oncogene product sufficient for association with phosphotyrosine-containing proteins. *Molecular and Cellular Biology*, **11**, 1607–13.

Musacchio, A., Gibson, T., Lehto, V.-P. and Saraste, M. (1992) SH3: an abundant protein domain in search of a function. *FEBS Letters*, **307**, 55–61.

Ren, R., Mayer, B. J., Cichetti, P. and Baltimore, D. (1993) Identification of a 10 amino acid proline rich SH3 binding site. *Science*, **259**, 1157–61.

Songyang, Z., Shoelson, S. E., Chaudhuri, M., *et al.* (1993) SH2 domains recognise specific phosphopeptide sequences. *Cell*, **72**, 767–78.

Oncogenes

Cantley, L. C., Auger, K. R., Carpernter, C., *et al.* (1991) Oncogenes and signal transduction. *Cell*, **64**, 281–302.

Watson, J. D., Gilman M., Witkowski, J. and Zoller, M. (1992) *Recombinant DNA* (2nd edn). W. H. Freeman, New York. Chapter 18 in particular.

History

Agranoff, B. W., Bradley, R. M. and Brady, R. O. (1958) The enzymatic synthesis of inositol phosphatide. *Journal of Biological Chemistry.*, **233**, 1077–83.

Ashman, D. F., Lipton, R., Melicow, M. M. and Price, T. D. (1963) Isolation of adenosine 3′,5′-monophosphate and guanosine 3′,5′-monophosphate from rat urine. *Biochemical and Biophysical Research Communications*, **11**, 330–4.

Hunnum, Y. A., Loomis, C. R., Merrill, A. H. and Bell, R. M. (1986) Sphingosine inhibition of protein kinase C and activity of phorbol dibutyrate binding *in vitro* and in human platelets. *Journal of Biological Chemistry.*, **261**, 2604–2609.

Hunter, T. and Sefton, B. M. (1980) Transforming gene product of Rous sarcoma virus phosphorylates tyrosine. *Proceedings of the National Academy of Sciences USA*, **77**, 1311–15.

Krupinski, J., Coussen, F., Bakalyar, H. A. *et al.* (1989) Adenylyl cyclase amino acid sequence: possible channel-like or transporter-like structure. *Science*, **244**, 1558-64.

Kuo, J. F. and Greengard, P. (1970) Cyclic nucleotide-dependent protein kinases. *Journal of Biological Chemistry*, **245**, 2493–8.

Paulus, H and Kennedy, E. P. (1960) The enzymatic synthesis of inositol monophosphate. *Journal of Biological Chemistry*, **235**, 1303–11.

Rall, T. W. and Sutherland, E. W. (1958) Formation of a cyclic adenine ribonucleotide by tissue particles. *Journal of Biological Chemistry*, **232**, 1065–76.

Sattin, A., Rall, T. W. and Zanella, J. (1975) Regulation of cyclic adenosine 3′,5′-monophosphate levels in guinea-pig cerebral cortex by interaction of α adrenergic and adenosine receptor activity. *Journal of Pharmacology and Experimental Therapeutics*, **192**, 22–32.

Sutherland, E. W. and Rall, T. W. (1957) The properties of an adenine ribonucleotide produced with cellular particles, ATP, Mg^{++}, and epinephrine or glucagon. *Journal of the American Chemical Society*, **79**, 3608.

Sutherland, E. W. and Rall, T. W. (1958) Fractionation and characterisation of a cyclic adenine ribonucleotide formed by tissue particles. *Journal of Biological Chemistry*, **232**, 1077–91.

Tonks, N. K., Diltz, C. D. and Fischer, E. H. (1988) Characterization of the major protein-tyrosine-phosphatases of human placenta. *Journal of Biological Chemistry*, **263**, 6731–7.

Whitman, M., Downes, C. P., Keeler, M., Keller, T. and Cantley, L. (1988) Type 1 phosphatidylinositol kinase makes a novel inositol phospholipid, phosphatidyl-inositol-3-phosphate. *Nature*, **332**, 644–6.

Techniques in the Study of Cell-signalling Components

Alberts, B., Bray, D., Lewis, J., Raff, M., Roberts, K. and Watson, J. D. (1994) *Molecular Biology of the Cell* (3rd edn). Garland Publishing, New York. In particular, pp. 778–82 for discussion on computer networks.

Hidaka, H. and Kobayashi, R. (1994) *Essays in Biochemistry*, **28**, 73–97. Protein kinase inhibitors.

Kendall, D. A. and Hill, S. J. (eds) (1995) *Signal Transduction Protocols*. Humana Press, New Jersey. A particularly useful collection of protocols for cell-signalling research.

Langdon, S. P. and Symth, J. F. (1995) Inhibition of cell signalling pathways. *Cancer Treatment Reviews*, **21**, 65–89.

Lin, L.-L., Lin, A. Y. and Knopf, J. L. (1992) Cytosolic phospholipase A_2 is coupled to hormonally regulated release of arachidonic acid. *Proceedings of the National Academy of Sciences USA*, **89**, 6147–51.

Rutter, G. A., White, M. R. H. and Tavaré, J. M. (1995) Involvement of MAP kinase in insulin signalling revealed by noninvasive imaging of luciferase gene-expression in single living cells. *Current Biology*, **5**, 890–9.

Van der Bend, R. L., Brunner, J., Jalink, K., Van Corven, E., Moolenaar, W. H. and Van Blitterswijk, W. J. (1992) Identification of a putative membrane-receptor for the bioactive phospholipid lysophosphatidic acid. *EMBO Journal*, **11**, 2495–501.

2 Extracellular Signals: Hormones, Cytokines and Growth Factors

Topics

- Hormones
- Plant Hormones
- Cytokines
- Growth Factors
- Neurotransmitters
- Pheromones
- ATP as an Extracellular Signal

Introduction

One of the most significant ways in which cells communicate with each other is by the release and detection of signalling molecules, such as hormones, cytokines and growth factors. Such molecules can be released some considerable distance from their point of action, and in general have an effect over a relatively long time-scale with an unspecific transport to the site of action. For example, a hormone might be released and carried by the bloodstream to all parts of the body where it washes around many types of cells, some of which will detect its presence. Many cells will be unaffected by such a release of hormone, the specificity of effect being determined by the presence of specific receptor molecules on the surface of the detecting cells.

Many of these extracellular signalling molecules are involved, and have been implicated, in disease states. Probably the most well studied and best understood is the role of insulin in diabetes. This multifactorial disease may involve aberrations in the synthesis of the hormone as well as its detection by the target cells. Insulin is further discussed in Chapter 9.

The role of hormones has been most widely studied in animals, although such extracellular signals are commonly used in the plant kingdom and have also been found to be important for unicellular eukaryotes, where the release of hormone-like molecules has been seen in the transfer of messages between individual organisms in a manner similar to transfer between tissues in mammals.

Although the distinction between hormones, cytokines and growth factors appears to be a somewhat loose one, a rough division is given here and they are considered separately.

Under the broad bracket of hormones are several diverse types of molecule. Although the exact definition of a hormone appears to be somewhat vague, here the term is used to describe substances released into the extracellular medium by the cells of one tissue and carried to a new site of action, such as a different tissue, where they provoke a specific response. They can be split into several broad classes: small water-soluble molecules; peptide hormones; lipophilic molecules detected by surface receptors; and lipophilic molecules detected by intracellular receptors.

Small Water-soluble Molecules

These hormones share common characteristics in that they are all water soluble and cannot cross the plasma membrane of the cell, usually because they carry a charge at physiological pH; therefore their detection is due to the presence of a specific receptor on the cell surface. Hormones that fall into this category include histamine and epinephrine (Figure 2.1).

Figure 2.1
Molecular structures of histamine and epinephrine, examples of small water-soluble hormones.

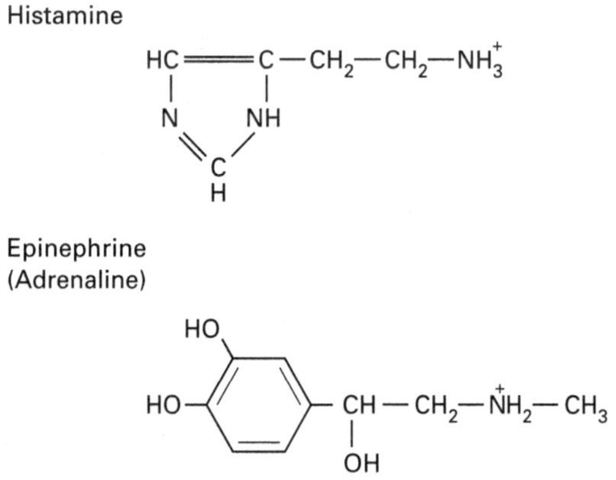

Histamine

Epinephrine
(Adrenaline)

Histamine is produced in the mast cells and is responsible for the control of blood-vessel dilation. Some topical drugs, for example those used for the treatment of bee stings, contain antihistamines, which are used to counteract the release of histamine during an inflammatory response. Epinephrine, or adrenaline as it used to be called, was classically known as the hormone released at times of panic, and has been referred to as the 'fight, fright and flight' hormone. It is produced by the adrenal medulla and causes an increase in blood pressure and pulse rate, contraction of smooth muscles, an increase in glycogen breakdown in muscles and the liver and an increase in lipid hydrolysis in adipose tissue, all factors that prepare the body for the situation where a greater energy requirement is envisaged, such as in fight and flight. In both cases the effects are fast and quickly reversed.

Peptide Hormones

Like histamine and epinephrine, the peptide hormones are water soluble and are carried to their site of action by the vascular system. Usually their release is rapid, as is their breakdown by proteases in the blood and tissues. The two that have been most extensively studied are glucagon and insulin, which in fact have antagonistic effects.

Insulin is produced by the β cells of the pancreas as a single proinsulin polypeptide (Figure 2.2). This folds into its correct secondary and tertiary structure aided and stabilised by the presence of three disulphide bridges. Once correctly folded, active insulin is produced by proteolytic cleavage in two places, causing removal of a substantial part of the middle of the polypeptide. The result is that insulin contains two polypeptide chains, an A chain of 21 amino acids and a B chain of 30 amino acids held together by disulphide bonds and electrostatic forces.

Figure 2.2
Production of insulin from proinsulin by the removal of the linking polypeptide (shown by the dotted line).

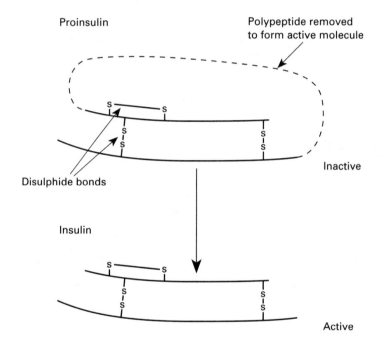

Once released, insulin stimulates the uptake of glucose by muscle and fat cells, increases lipid synthesis in adipose tissues and causes a general increase in cell proliferation and protein synthesis. Detection of insulin is due to the presence on the surface of target cells of a multipolypeptide receptor, which contains intrinsic tyrosine kinase activity. Activation of the receptor leads to a cascade of events including the activation of phosphatidylinositol 3-kinase, activation of G proteins and the stimulation of a mitogen-activated protein (MAP) kinase pathway. These events can lead to the stimulation of transcription factors and increase in selected gene expression, as discussed further in Chapter 9.

Glucagon, like insulin and all peptide hormones, is also derived from a precursor polypeptide. Active glucagon contains a single polypeptide chain of 29 amino acids. It is released from the α cells of the pancreas and leads to glycogen breakdown and lipid hydrolysis, allowing an increase in glycolysis and respiratory rates.

Other hormones in this class include follicle-stimulating hormone (FSH) and luteinising hormone (LH). Both are like insulin in having two polypeptide chains. However, they are much bigger, FSH having an α-chain of 92 amino acids with a β-chain of 118 amino acids. Both LH and FSH are produced by the anterior pituitary. FSH stimulates the growth of ovarian follicles and oocytes and also increases the production of oestrogen, while LH controls the maturation of oocytes and increases the release of oestrogen and progesterone. The release of LH is also under the control of a peptide hormone, this time a single polypeptide hormone called LH-releasing hormone produced by the hypothalamus and neurones. This type of interaction, i.e. different hormones controlling each other's synthesis and secretion, is not unusual.

Lipophilic Molecules Detected by Cell-surface Receptors

Receptors on the surface of cells also detect the presence of a group of hydrophobic or lipophilic molecules that act like hormones. The main signalling molecules in this context are prostaglandins. These are synthesised from arachidonic acid, a 20-carbon fatty acid that can be hydrolytically released from the lipid plasma membrane. Although they are a relatively large group of chemicals they can be roughly divided into nine different classes. Although originally named after the organ where they were thought to be produced, prostaglandins are in fact produced by most cells and their action is usually local. Their synthesis can be inhibited by several anti-inflammatory drugs including aspirin. Their effects range from control of platelet aggregation to the induction of uterine contraction.

Lipophilic Molecules Detected by Intracellular Receptors

This class of hormones encompasses the steroid hormones, for example, oestrogen, progesterone, androgens and glucocorticoids, as well as the thyroid hormones and retinoids (Figure 2.3).

Steroid hormones are all derived from cholesterol and are synthesised and secreted by endocrine cells. Progesterone is synthesised by the ovaries and placenta and is involved in the development of the uterus in readiness for implantation of the new embryo, as well as for the stabilisation of early pregnancy and development of the mammary glands. Similarly, testosterone produced by the testis is responsible for the development and functioning of the male sex organs, as well as the development of the male characteristics such as hair growth. Oestrogens such as oestradiol are also involved in the development of female sexual characteristics such as uterine differentiation and mammary gland function. In insects and crustaceans, a similar role in the development of sexual characteristics is fulfilled by a steroid-like compound called α-ecdysone.

Figure 2.3
Molecular structures of testosterone, progesterone and thyroxine, examples of lipophilic molecules detected by intracellular receptors.

Testosterone

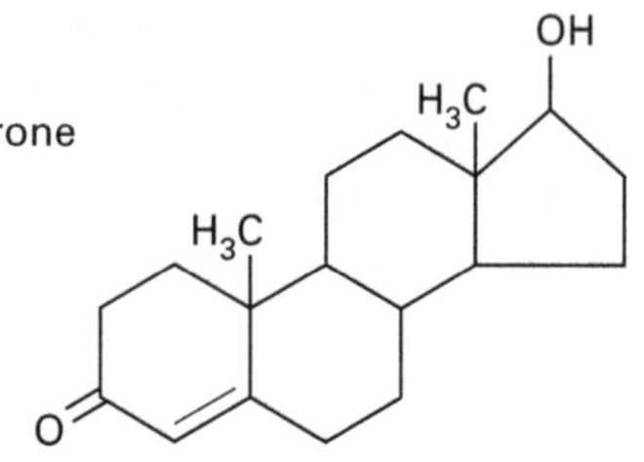

Progesterone

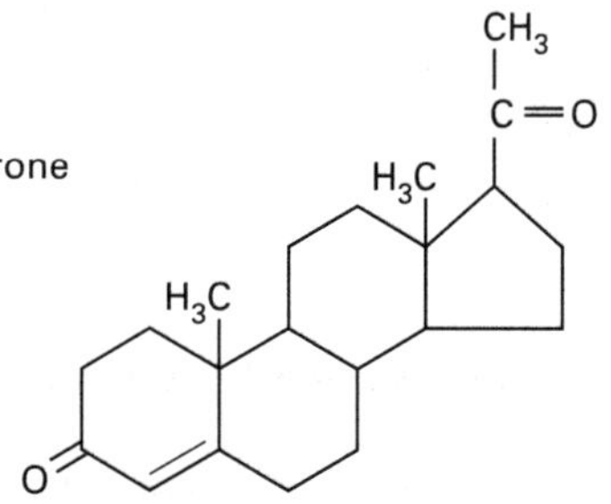

Thyroxine
(Tetraiodothyronine)

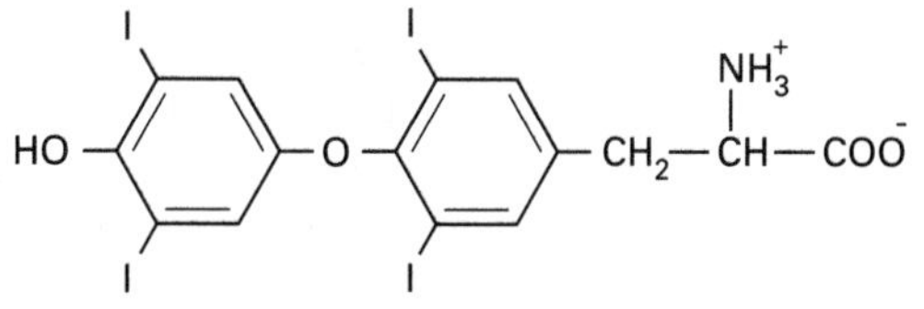

Other steroid hormones include cortisol and vitamin D. Cortisol, a glucocorticoid, is produced by the cells of the adrenal cortex and controls the metabolic rates of many cells. It is formed from progesterone by three hydroxylation steps, at carbons 11, 17 and 21. Vitamin D is synthesised in the skin exposed to sunlight: 7-dehydrocholesterol (provitamin D_3) in the skin is lysed by ultraviolet (UV) light to previtamin D_3, an inactive form, which is activated by hydroxylation in the liver and kidneys with conversion to calcitriol (1,25-dihydroxycholecalciferol). Calcitriol is responsible for the control of Ca^{2+} uptake in the gut and for decreasing Ca^{2+} excretion by the kidneys. Lack of vitamin D in a child's development can lead to the development of rickets, a condition in which the cartilage and bone fail to calcify properly leading to malformation of the bones, especially the long bones. This is most noticeable in the legs of patients and was not uncommon when children were forced to work long hours indoors or underground away from adequate sunlight. Dietary vitamin D, such as derived from fish oils, overcomes the lack of its *de novo* synthesis in the skin and prevents the associated symptoms.

Thyroid hormones are derived from the amino acid tyrosine. Thyroxine, otherwise known as tetraiodothyronine, is produced by the thyroid gland and has effects on the metabolism of many cells, including increase in heat production and production of polypeptides involved in metabolic pathways.

Retinoids are synthesised from vitamin A, also called retinol. Retinoic acid is formed by the oxidation of the alcohol group of retinol to a carboxyl group. Binding of this molecule to its receptor can lead to an alteration of gene expression. Retinol is also the precursor of retinal, the light-sensitive group of rhodopsin. It is thought that a deficiency in vitamin A can lead to an impairment in light detection by the eye under low lighting conditions, a disorder commonly called night blindness.

Because these lipophilic molecules are inherently insoluble in water, once released into the bloodstream they are stabilised by association with specific carrier proteins, from which they dissociate before they cross the plasma membrane and enter the cells. In general, these groups of hormones can persist in the circulation for hours or even days and are involved in long-term control. However, once inside the cells they are detected by receptors in the cytosol or the nucleus, often culminating in the alteration of specific gene expression.

Plant Hormones

The existence of plant hormones was first postulated by Darwin who showed by experimenting on the phototropism of coleoptiles that substances must be transported in plants. By 1928 Went had discovered auxin and now a wide range of substances have been placed under the umbrella of plant hormones. It is in fact a very loose term and other terms, such as 'plant growth substance' and 'phytohormones', have been suggested but never used extensively. In animals, extracellular signalling molecules have been divided, albeit rather crudely, into classes such as hormones and cytokines. The term 'plant hormone' embraces all the substances that act as extracellular signals in plants (Figure 2.4), which means that they might exert their influence at a location some distance from their site of manufacture, or might act within the tissue where they are made or even within the same cell. For example, ethylene has effects over a very short distance, while cytokinins can be transported from the roots to the leaves.

Auxin

Although the main auxin found in plants is indole-3-acetic acid (IAA; Figure 2.4), several compounds derived from this molecule show auxin-like activity, including indoleacetaldehyde or indoleacetyl aspartate. IAA is manufactured from tryptophan or indole in the leaves, particularly young leaves, and developing seeds. Its transport appears to be primarily from cell to cell but it can also be found in the phloem. Its effects are diverse, ranging from the stimulation of cell enlargement and stem growth, stimulation of cell division and the differentiation of the phloem and xylem vessels to the mediation of tropic responses to light and gravity.

Figure 2.4
Molecular structures of auxin, zeatin, abscisic acid and ethylene, examples of plant hormones.

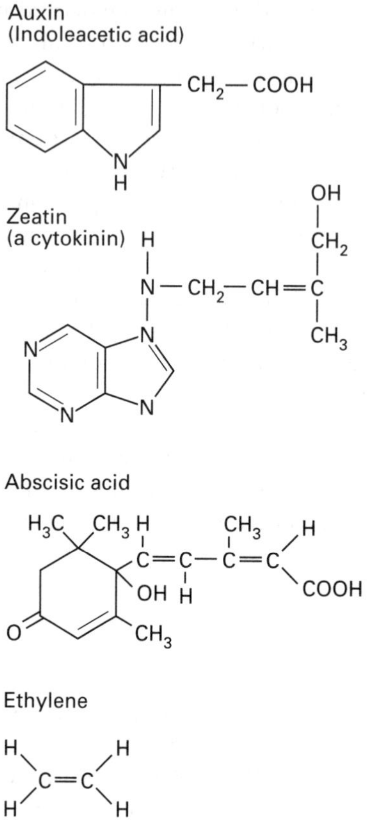

Auxin
(Indoleacetic acid)

Zeatin
(a cytokinin)

Abscisic acid

Ethylene

Cytokinins

Cytokinins (CKs) are derived from adenine in the root tips and developing seeds, the most common being zeatin (Figure 2.4). Their transport is via the xylem system and their activities include the induction of cell division, although this requires the presence of auxin, leaf expansion caused by cell enlargement and delay of leaf senescence.

Gibberellins

The most common biologically active gibberellin in plants is gibberellin A_1 (GA_1), although the gibberellins form a family of active compounds. Their structure, known as the *ent*-gibberellane structure, is complex and involves four ring structures bridged by a fifth oxygen-containing ring. They are synthesised from mevalonic acid in young shoots and developing seeds and their transport is via both the xylem and phloem. Their biological activities include the stimulation of flowering and seed germination, stimulation of the synthesis of enzymes such as α-amylase and stimulation of cell division and cell elongations, which manifests itself as elongation of the stems.

Abscisic Acid

Like the gibberellins, abscisic acid (ABA) is synthesised from mevalonic acid, but this time mainly in the roots and mature leaves, although most tissues are probably capable of producing it. Seeds can also synthesise ABA or indeed import it from the parent plant. ABA is transported by the phloem down to the roots and by the xylem up to the leaves. It is synthesised particularly in times of water stress and one of the functions of ABA is to mediate stomatal closure. It also induces the synthesis of storage proteins in seeds and inhibits shoot growth.

Ethylene

What appears to be an odd compound to act as a signalling molecule is the gas ethylene (Figure 2.4). However, analogous compounds have been discovered in animals, as discussed in Chapter 8. Ethylene is synthesised from methionine in most plant tissues, particularly in response to stress. Its transport is by diffusion, although intermediates in its synthesis can be transported around the plant and lead to its appearance at a site distant from the original stimulus. One of the most commonly known roles for ethylene is in the induction of fruit ripening but it can also lead to shoot and root differentiation and growth and to opening of flowers, amongst many other effects.

Other Plant Hormones

Many other compounds come under this broad term of plant hormone, including jasmonates (discussed in Chapter 6), polyamines, salicylic acids and brassinosteroids. This last group contains over 60 steroidal compounds, while salicylic acid has been implicated in thermogenesis and the production of pathogen-induced proteins.

Cytokines

Cytokines are a group of peptide molecules produced by cells and have their effect on other cells within a short distance or even act on themselves. Hence, cytokine effects tend to be local, i.e. they are involved in paracrine or autocrine functions. Under this definition are those molecules principally responsible for the coordination of the immune response of higher animals and include the interleukin series, tumour necrosis factors and interferons. The number of molecules found to belong to this group appears to be increasing rapidly, with regular reports of the discovery of a new interleukin, the number having risen from 8 to 17 in a matter of 4–5 years.

Interleukins

The interleukin (IL) series has 17 members designated IL-1 to IL-17, although this will undoubtedly increase rapidly with the employment of modern molecular biological techniques. However, some of these interleukins are themselves subdivided into separate groups. For example, IL-1 exists in two distinct forms, IL-1α and IL-1β, which are coded for by separate genes. Both are produced as a larger precursor molecule and cleavage produces the active extracellular form. IL-1α is a 159-amino acid peptide derived from a 271-amino acid precursor, while IL-1β is a 153-amino acid peptide derived from a 269-amino acid precursor. Therefore, similar to the peptide hormones such as insulin, a major cleavage event has taken place that removes a substantial proportion of the polypeptide. Such cleavage events are common in the synthesis of the interleukin family. In the case of IL-1, the uncleaved cell-associated form seems to retain biological activity. Several of the genes for the cytokines, for example IL-3, IL-4, IL-5, IL-13 along with granulocyte–macrophage colony-stimulating factor (GM-CSF), are grouped together on the same region of chromosome 5 in humans or chromosome 11 in mice, suggesting that they originally arose through gene duplication. Table 2.1 summarises some of the characteristics of the interleukins.

Table 2.1. The interleukin family and some of their characteristics.

| | Amino acids | | Chromosomal | |
Cytokine	Percursor	Active form	location	Notes
IL-1α	271	159	2q13	
IL-1β	269	153	2q13–q21	
IL-2	153	133	4q26–q27	
IL-3	152	134/140	5q23–q31	Chromosome 11 mouse
IL-4	153	129	5q23–q32	Chromosome 11 mouse
IL-5	134	115	5q23.3–q32	Chromosome 11 mouse: homodimer
IL-6	212	184	7p15–p21	
IL-7	177	152		
IL-8	99	72–77		Homodimer
IL-9	144	126	5q31–q35	
IL-10	178	160	1	Homodimer: total 35 kDA
IL-11	199	178	19q13.3–13.4	
IL-12		Total 75 kDa		Heterodimer, p35 + p40
IL-13		132	5q31	Chromosome 11 mouse
IL-14		53 kDa		
IL-15	162	114		

Most of the interleukins are active in the monomeric state, but IL-5 exists as a homodimer of two 115-amino acid polypeptides, while IL-12 exists as a heterodimer of variable-sized polypeptides. IL-8, otherwise known as neutrophil-activating protein (NAP-1), is also a homodimer but here the subunits are of variable length (72–77 amino acids) due to truncation of the N-terminal end of the polypeptides. It is interesting that the IL-8 molecules which contain the shorter versions of the subunits are more potent than the ones with the longer subunits.

Cytokines are produced by a variety of blood cells and cells involved in the immune response. For example, IL-2 and IL-3 are produced mainly by helper T lymphocytes, while IL-6 is produced by a variety of cells including T cells, macrophages and fibroblasts.

The biological responses of the interleukin family are also varied. IL-13, formerly known as P600, seems to induce the growth and differentiation of B cells and inhibits cytokine production by macrophages and their precursor cells, monocytes. IL-5 leads to the activation of eosinophil function, including chemotaxis and eosinophil differentiation. Some of the individual cytokines have a diverse range of biological activities, a good example being IL-6. It stimulates the differentiation of myeloid cell lines, acts as a growth factor on B cells, modulates the responses of stem cells to other cytokines and has effects on non-haematopoietic cells, including the development of nerve cells.

These cytokines all work in a concerted way to coordinate the whole response. They can exert coordinating responses on the cells themselves or regulate the synthesis of each other. For example, IL-5 enhances the IL-4 effect on B cells, increasing the IL-4-induced synthesis of IgE and the expression of CD23, while IL-11 has synergistic effects with both IL-3 and IL-4. On the other hand, IL-6 induces the production of IL-2 in T cells but IL-10 inhibits the synthesis of several other cytokines, including IL-1, IL-6, IL-8, IL-10 and IL-12. Some cytokines produce similar responses. In the case of IL-13 and IL-4 no additive effect is seen if both of these cytokines are added together, and it is possible that they share a common receptor or share the same signal-transduction pathway. Likewise, IL-15 and IL-2 may share common elements in their receptors.

An interesting interleukin is IL-1ra, which acts as an antagonist to IL-1 by competing for receptor-binding sites, so reducing the effect of IL-1. The production of IL-1ra may itself be under the control of IL-13.

Interferons

The interferons (IFNs) fall into two main groups, α/β and γ. IFN-γ rose to prominence in the press as an anticancer agent and was branded as a wonder drug. Like the interleukins, IFN-γ is a peptide: it is 143 amino acids long and exists as a dimer in the active form. However, it is synthesised from the expression of a single gene on the long arm of chromosome 12 in humans, the sequence including a signal sequence of 23 amino acids. The α/β forms, on the other hand, are monomers, coded for by genes on chromosome 9.

Produced by T cells amongst others, the effects of IFN-γ include the induction of expression of class II histocompatibility antigens on epithelial, endothelial and connective tissue. It also acts as a macrophage-activating factor, causing specific gene expression in these cells that leads to the enhanced cytotoxic activity against tumours and enhanced killing of parasites.

Tumour Necrosis Factors

Tumour necrosis factor α, (TNF-α), originally known as cachectin, is coded for by a single gene, which in humans is found on chromosome 6. In the active form TNF-α is a trimer of 157-amino acid subunits. It is produced by monocytes and macrophages. Its activity is closely coupled to that of IFN-γ and it has been shown to cause necrosis of tumours, hence its name, and to be cytotoxic to transformed cells *in vitro*. The gene for TNF-β is close to the TNF-α gene, both having similar biological activity and even binding to the same receptor.

Other Cytokines, Chemokines and Receptors

There are many other cytokines, including granulocyte colony-stimulating factor (G-CSF) and GM-CSF, but it is beyond the scope of this book to cover them all.

Many of the small protein signalling molecules come under the grouping of the chemokines. This is a large family, or superfamily, of peptides grouped together because of their structural relationships rather than any functional similarity. There are two sub-groups of the chemokine family, characterised by a particular structural motif containing four cysteine residues. The first sub-group is known as the C-C chemokines and contains the sequence:

N terminus-X(10–11)-Cys-Cys-X(22–23)-Cys-X(15)-Cys-X(18–24)-C terminus

where the numbers in parentheses denote the number of amino acid residues likely to be present at this point in the primary structure. The second subgroup, termed the C-X-C chemokines, contains the sequence:

N terminus-X(6–12)-Cys-X(1)-Cys-X(23–24)-Cys-X(15–16)-Cys-X(15–53)-C terminus

where the first two cysteine residues are separated by one amino acid. Chemokines have been alternatively referred to by the names intercrines, the small cytokine family (scy) and the small inducible secreted cytokine (SIS). It should also be noted that this group contains cytokines grouped elsewhere, such as IL-8 discussed above. Many of these chemokines have been discovered not by their functional activity but by their sequence homology when their cDNAs have been cloned. At least 20 separate proteins could be listed here and the number is sure to rise.

Many of the receptors for cytokines have been found to contain some similarity in their structure and primary amino acid sequence: four cysteine residues amongst a hydrophobic region together with a hydrophobic region encompassing runs of positively and then negatively charged residues have been found in several of the receptors cloned. On binding of the cytokine intracellular signalling usually involves tyrosine phosphorylation, as discussed in Chapter 4.

The term 'growth factor' is used here to define those compounds shown to have specific functions in the regulation of the growth and differentiation of cells. Other works may well categorise members of this group under the other headings used above, and this emphasises the vagueness of the terms used. Approximately 50 proteins that possess growth factor-like activity have been reported, with at least 14 different receptor families involved in their detection.

Platelet-derived Growth Factor

PDGF is a dimeric protein that may contain two related polypeptides, A or B. The active growth factor has a molecular mass of between 28 and 35 kDa, and is made up of two of the same subunits, for example AA or BB, or may be a heterodimeric protein, AB. The individual subunits have a molecular mass of 12–18 kDa and are coded for by two separate genes. They are held together by disulphide bonds, and in fact all the cysteine residues in the polypeptides are involved in either intermolecular or intramolecular bonding. Interestingly, the receptor for PDGF is also dimeric: it is a homodimer of two identical subunits, $\alpha\alpha$, or $\beta\beta$, or is a heterodimer, $\alpha\beta$. The different forms of PDGF. PDGF-BB, with two B subunits, can bind and activate all the receptor forms, whereas PDGF-AA, with two A subunits, can only activate the $\alpha\alpha$ receptor. The heterodimeric PDGF-AB has intermediate activity.

PDGF has been shown to induce both cell migration and cell proliferation and has been implicated in several disease states, including fibrosis and arteriosclerosis.

Epidermal Growth Factor

A larger group of growth factors comes under the heading of EGF. These include EGF itself, transforming growth factor α (TGF-α), betacellulin and heparin-binding EGF. They are very small peptides, rat EGF being only 5.2 kDa, although their precursors are very large. The EGF precursor is in fact a transmembrane protein of 1168 amino acids, of which only 53 are cleaved off to become the active EGF molecule.

These growth factors are characterised by having several aromatic residues exposed to the aqueous medium. This is unusual in proteins, which in general try to hide these hydrophobic side-chains in the interior of the folded polypeptide structure. It has been proposed that there are interactions between these aromatic groups, and the protein surface may contain an aromatic domain vital for its function. Proteins with domains related in sequence to EGF have been found in *Drosophila* and sea-urchin embryos. The proposed signal-transduction cascade leading from EGF receptors is discussed in Chapter 4 and illustrated by Figure 4.9.

Fibroblast Growth Factor

Fibroblast growth factor (FGF) represents a family of molecules involved in the regulation of proliferation, differentiation and cell mobility. In mammals, the family comprises nine structurally related proteins of 20–30 kDa. The nine genes probably arose from gene duplication of an ancestral gene, but further isoforms can arise from alternative splicing or the use of alternative initiation codons for translation. Alternative post-translational modification leads to further diversification of the molecules within the family, so that some are glycosylated, phosphorylated or even methylated or cleaved.

Some members of the FGF family have been identified as oncogenes, causing the transformation of cells and uncontrolled proliferation leading to the formation of tumours.

Neurotransmitters

In the axon termini of presynaptic cells of the nervous system there are storage vesicles called synaptic vesicles that contain neurotransmitters. Voltage-gated Ca^{2+} channels are opened on the arrival of an action potential, leading to a sharp increase in the concentration of intracellular Ca^{2+}. This in turn leads to exocytosis from the synaptic vesicles, releasing the neurotransmitters into the space between the nerve cells. Receptors on the postsynaptic cells detect the presence of these compounds, leading to the propagation of the signal or the response.

These transmitters fall into two main groups. The first group is composed of a group of small molecules and contains some classical neurotransmitters, including acetylcholine, γ-aminobutyric acid (GABA) and dopamine (Figure 2.5). This group also includes some molecules already discussed, such as epinephrine and histamine.

Figure 2.5
Molecular structures of acetylcholine, dopamine and GABA, examples of neurotransmitters.

Acetylcholine

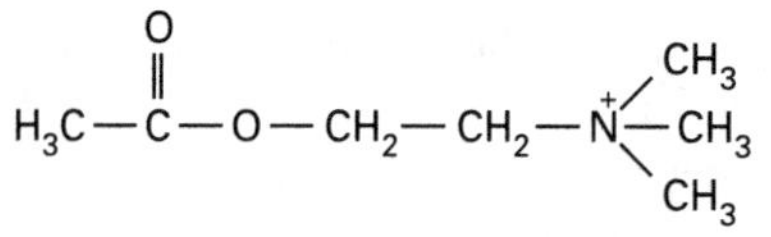

Dopamine

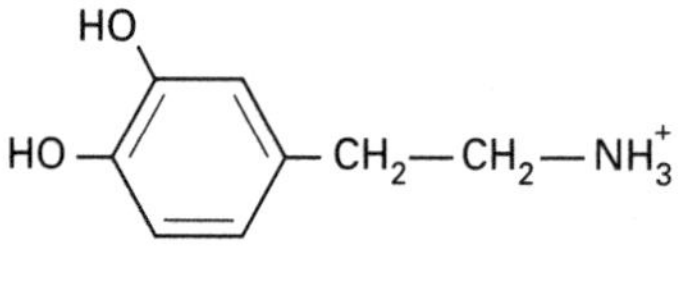

γ-Aminobutyric acid (GABA)

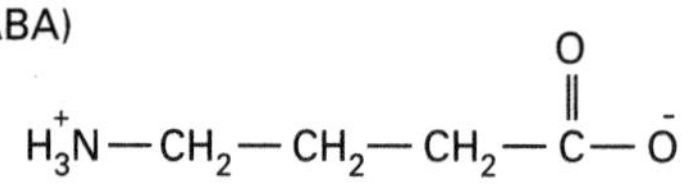

Also included here are amino acids, such as glycine and glutamine, and derivatives of amino acids, e.g. dopamine is derived from tyrosine, serotonin is a derivative of tryptophan, while GABA is derived from glutamate. Also included are nucleotide-derived compounds such as adenosine triphosphate (ATP) and adenosine.

The second group are the neuropeptides, including vasopressin, bradykinin and adrenocorticotrophic hormone.

The effects of neurotransmitters are very local, acting on the cell across the synaptic space or at least within a short distance. The receptors on the postsynaptic cell fall into two classes, either having an intracellular effect through a G protein, as seen for muscarinic receptors binding to acetylcholine, or acting as ligand-gated ion channels. If the ion-gated channel has a specificity for Na^+/K^+ then the resulting response is excitatory, but if the receptor specificity is for Cl^- then an inhibitory response results.

Pheromones

Karlson and Lüscher first defined pheromones in 1959. They are substances excreted by an individual and which have their effect on another individual of the same species. The word 'pheromone' comes from the Greek *pherein*, which means to transfer, and *hormon*, which means to excite.

Many bacterial species use pheromones in communication, the effects usually being seen if the cells grow to a particular density. Responses induced by pheromones include the production of light (luminescence), the production of virulence factors, the development of fruiting bodies and plasmid transfer.

The chemical structure of bacterial pheromones seems to be quite variant and includes amino acids, short peptides, proteins and branched-chain fatty acids. One of the largest group of pheromones are the so-called *N*-acetyl-L-homoserine lactones (AHLs). Many AHLs not only induce a response in another individual but also induce the genes for their own production. The transcriptional activators used in the AHL response are grouped together in what is called the LuxR superfamily of response regulators, of which at least 15 members are known. Similarly, a superfamily of enzymes that produce AHLs have been defined, the 10 members of this family being known as the LuxI superfamily.

It is not just prokaryotes that use pheromones in their organism-to-organism communication; such systems are used extensively by higher organisms. The so-called water mould, *Allomyces*, uses a compound called sirein as a sexual attractant. This compound is related to a cyclic organic compound called 2-carene, which is found in pine resins. Another slime mould, *Achlya*, uses two steroid pheromones, one produced by the male and one produced by the female. Detection of male pheromone by the female is essential for the development of the female's sexual machinery, and likewise the male needs to detect the female's pheromone.

The sea anemone *Anthopleura elegantissima* uses a positively charged organic compound called anthopleurine. This is an interesting pheromone as it is distributed by a second species. The sea anemones are eaten by a sea slug and thus the sea slug ingests the pheromone. The pheromone is then released by the sea slug as it moves around and this acts as a warning to other sea anemones that a predator is coming. The sea anemones near the advancing sea slug will then retract as a defence.

Much of the early work on pheromones was carried out by Adolph Butenandt, a German organic chemist. He worked with the silkworm moth, *Bombyx mori*. This insect uses a compound that has been named bombykol, a long 16-carbon unsaturated fatty acid. Much work has been done subsequently to study the pheromones of other insects.

Higher organisms also use pheromones in their silent communications, including fish, amphibians and even mammals, but it is impossible to discuss them all here.

ATP as an Extracellular Signal

Extracellular ATP can arise through release from cells involved in the secretion of neurotransmitters or from storage granules in cells such as adrenal medulla cells or lymphocytes. Cell death and breakage can also lead to the release of intracellular ATP into the extracellular medium and therefore the local concentrations of ATP may be in the nanomolar or even micromolar range. However, three different ectonucleotidases are responsible for the sequential hydrolysis of ATP to adenosine, so removing ATP from solution. Even so, many cells have been shown to possess purinoceptors on their surface capable of detecting ATP, including platelets, neutrophils, fibroblasts, smooth muscle cells and cells of the pancreas. These receptors fall into two groups, P_1 and P_2, depending on their specificity for the adenosine compound, P_2 receptors having the highest affinity for ATP. P_2 receptors can be subdivided into four groups depending on their action, some acting through G proteins with a commensurate effect on phosphatidylinositols via PLC, while others act through the operation of cation channels.

In a similar fashion, the slime mould *Dictyostelium discoideum* can exploit the adenosine compound cAMP as an extracellular signal, controlling differentiation and cellular aggregation. If a food source becomes scarce, cAMP is used as a chemo-attractant to signal to the normally free-swimming cells to aggregate into a slug, accompanied also by changes in gene expression.

Summary

Cells commonly communicate by the release and detection of signalling molecules, often over relatively large distances. Classically, these compounds have been referred to as hormones. More recently, other molecules have been categorised separately and are known as cytokines, chemokines or growth factors.

Hormones are a diverse group of molecules, which may be small and water soluble, peptides or lipophilic molecules, the latter being detected by either a cell-surface receptor or an intracellular receptor, as seen with steroid hormones.

Plant hormones too are a very diverse group of chemicals, including compounds derived from amino acids, lipids or even a gas, such as ethylene.

Cytokines are an expanding group of peptides that include the interleukins, currently having 17 members, the interferons and tumour necrosis factors. Such molecules are important in the orchestration of the immune response and the development of cells used in animal host defence.

Growth factors are molecules involved in the regulation of the growth and differentiation of cells. The family is known to contain at least 50 proteins but, as with the cytokines and chemokines, undoubtedly more are yet to be discovered.

Other more unusual extracellular signals are thought to include the energy storage compound ATP and a compound usually associated with intracellular signalling, cAMP, as seen in control of the aggregation of the slime mould *Dictyostelium*.

Further Reading

Introduction

Baulieu, E. E. and Kelly, P. A. (1990) *Hormones: from Molecules to Disease*. Chapman & Hall, London.

Le Roith, D., Shiloach, J., Roth, J. and Lesniak, M. A. (1980) Evolutionary origins of vertebrate hormones: substances similar to mammalian insulins are native to unicellular eukaryotes. *Proceedings of the National Academy of Sciences USA*, **77**, 6184–8.

Turner, A. J. (ed.) (1994) *Neuropeptide Gene Expression*. Portland Press, London.

Hormones

DeLuca, H. F. (1992) New concepts of vitamin D functions. *Annals of the New York Academy of Sciences*, **669**, 59–68.

Jung, L. J., Kreiner, T. and Scheller, R. H. (1993) Prohormone structure governs proteolytic processing and sorting in the Golgi complex. *Recent Progress in Hormone Research*, **48**, 415–36.

Orci, L., Ravazzola, M., Storch, M.-J., Anderson, R. G. W., Vassalli, J.-D. and Perrelet, A. (1987) Proteolytic maturation of insulin is a post-Golgi event which occurs in acidifying clathrin-coated secretory vesicles. *Cell*, **49**, 865–8.

Steiner, D. F., Smeekens, S. P., Ohagi, S. and Chan, S. J. (1992) The new enzymology of precursor processing endoproteases. *Journal of Biological Chemistry*, **267**, 23435–8.

Cytokines

Ambrus, J. L., Pippin, J., Joseph, A. *et al.* Identification of a cDNA for a human high-molecular-weight B-cell growth factor. *Proceedings of the National Academy of Sciences USA*, **90**, 6330–4.

Arai, K.-I., Lee, F., Miyajima, A., Miyatake, S., Arai, N. and Yokota, T. (1990) Cytokines: coordination of immune and inflammatory responses. *Annual Review of Biochemistry*, **59**, 783–836.

Beutler, B. and Cerami, A. (1989) The biology of cachectin/TNF: a primary mediator of the host response. *Annual Review of Immunology*, **7**, 625–55.

de Waal Malefyt, R., Figdor, C. G., Huijbens, R., *et al.* (1993) Effects of IL-13 on phenotype, cytokine production and cytotoxic function of human monocytes: comparison with IL-4 and modulation by IFN-γ or IL-10. *Journal of Immunology*, **151**, 6370–81.

Dinarello, C. A. (1991) Interleukin 1 and interleukin 1 antagonism. *Blood*, **77**, 1627–52.

Farrar, M. and Shreider, R. D. (1993) The molecular cell biology of interferon-γ and its receptor. *Annual Review of Immunology*, **11**, 571–611.

Grabstein, K. J., Eisenman, J., Shanebeck, K., *et al.* (1994) Cloning of a T cell growth factor that interacts with the β-chain of the interleukin 2 receptor. *Science,* **264**, 965–8.

Gubler, U., Chua, A. O., Gately, M. K., *et al.* (1991) Regulation of human lymphocyte proliferation by a heterodimeric cytokine, IL-12 (cytotoxic lymphocyte maturation factor). *Proceedings of the National Academy of Sciences USA*, **88**, 4142–7.

Kurt-Jones, E. A., Kiely, J. M. and Unanue, E. R. (1985) Conditions required for expression of membrane IL-1 on B cells. *Journal of Immunology*, **135**, 1548–50.

McKenzie, A. N. J., Culpepper, J. A., de Waal Malefyt, R. *et al.* (1993) Interleukin 13: a T cell derived cytokine that regulates human monocyte and B cell function. *Proceedings of the National Academy of Sciences USA*, **90**, 3735–9.

McKenzie, A. N. J., Li, X., Largaespada, D. A., *et al.* (1993) Structural comparison and chromosomal localisation of the human and mouse IL-13 genes. *Journal of Immunology*, **150**, 5436–5444.

Minami, Y., Kono, T., Miyazaki, T. and Taniguchi, T. (1993) The IL-2 receptor complex: its structure, function and target genes. *Annual Review of Immunology*, **11**, 245–67.

Moore, K., O'Garra, A., de Waal Malefyt, R., Vierra, P. and Mosmann, T. (1993) Interleukin-10. *Annual Review of Immunology*, **11**, 165–90.

Scott, P. (1993) IL-12: initiation cytokine for cell mediated immunity. *Science*, **260**, 496–7.

Thomson, A. (ed.), (1994) *The Cytokine Handbook* (2nd edn). Academic Press, London. An excellent up-to-date review of the area.

Wolf, S. F., Temple, P. A., Kobayashi, M. *et al.* (1991) Cloning of cDNA for natural killer cell stimulatory factor: a heterodimeric cytokine with multiple biological effects on T cells and natural killer cells. *Journal of Immunology*, **146**, 3074–81.

Growth Factors

Bradshaw, R. A., Blundell, T. L., Lapatto, R., McDonald, N. Q. and Muray-Rust, J. L. (1993) Nerve growth factor revisited. *Trends in Biochemical Sciences*, **18**, 48–52.

Johnson, A., Heldin, C.-H., Westmark, B. and Wasteson, A. (1982) Platelet–derived growth factor: identification of constituent polypeptide chains. *Biochemical and Biophysical Research Communications*, **104**, 66–71.

Mason, I. J. (1994) The ins and outs of fibroblast growth factor. *Cell*, **78**, 547–552.

Mayo, K. H., Schaudies, P., Savage, C. R., De Marco, A. and Kapten, R. (1986) Structural characteristics and exposure of aromatic residues in epidermal growth factor from the rat. *Biochemical Journal*, **239**, 13–18.

Oefner, C., D'Arcy, A., Winkler, F. K., Eggimann, B. and Hosang M. (1992) Crystal structure of human platelet-derived growth factor BB. *EMBO Journal*, **11**, 3921–6.

Yang, Y., Ricciardi, S., Ciarletta, A., Kelleher, K. and Clark, S.C. (1989) Expression cloning of a cDNA encoding a novel human hematopoietic growth factor: human homologue of murine T-cell growth factor P40. *Blood*, **74**, 1880–4.

Pheromones

Agosta, W. C. (1992) *Chemical Communications: the Language of Pheromones*. W. H. Freeman, New York.

Kell, D. B., Kaprelyants, A. S. and Grafen, A. (1995) Pheromones, social behavior and the functions of secondary metabolism in bacteria. *Trends in Ecology and Evolution*, **10**, 126–9.

Wirth, R., Muscholl, A. and Wanner, G. (1996) The role of pheromones in bacterial interactions. *Trends in Microbiology*, **96**, 96–103.

ATP as an Extracellular Signal

Burnstock, G. (1978) A basis for distinguishing two types of purinergic receptor. *Cell Membrane Receptors for Drugs and Hormones: a Multidisciplinary Approach*, Straub, R. W. and Bolis, L. (eds.), pp. 107–18. Raven Press, New York.

El-Moatassim, C., Dornand, J. and Mani, J.-C. (1992) Extracellular ATP and cell signalling. *Biochimica et Biophysica Acta*, **1134**, 31–45.

Gordon, J. L. (1986) Extracellular ATP: effects, sources and fate. *Biochemical Journal*, **233**, 309–19.

3 Detection of Extracellular Signals: Receptors

Topics

- Types of Receptors
- Ligand Binding
- Receptor Sensitivity and Receptor Density

Types of Receptors

As discussed in Chapter 2, many signals reach a cell from the extracellular environment, whether via diffusion in the growth medium or actively through a vascular system. Usually extracellular molecular signals are found at very low concentrations, in the order of 10^{-8} mol/l. If the cell is to respond to the presence of such a signal, the cell must have the capacity to detect the signal molecule and to act upon it. Detection of the signal is usually accomplished by a specific receptor on the cell surface, where the receptor has a high binding affinity in the concentration range of that of the ligand. Binding of the ligand to this receptor will then stimulate the required intracellular response usually via a complex signal-transduction pathway. The detection of the signalling molecule has to be precise and the cell has to have the ability to respond repeatedly to the same signal or simultaneously to several signals. Therefore, a variety of receptors have evolved to fill this vital role. However, despite the vast array of extracellular molecules that need to be detected by a single cell, receptors fall mainly into five classes: ion channel linked; G protein linked; containing intrinsic enzymatic activity; tyrosine kinase linked; intracellular (Figure 3.1).

G Protein-linked Receptors

This class of receptor, when activated by binding to its ligand, results in the activation of a G protein, which conveys the message to the next component in the signal pathway. These receptors include those which have a specificity for hormones such as epinephrine, serotonin and glucagon along with those listed in Table 3.1. These receptors are of great interest to the pharmaceutical industry as they can be targets of therapeutic interest. For example, drugs that act via this pathway include some antihistamines, anticholinergics, inhibitors of the β-adrenergic receptor and some opiates.

Figure 3.1
Three different types of cell receptor: (a) ion channel–linked receptor, (b) G protein–linked receptor; (c) receptor containing intrinsic enzyme activity.

(a)

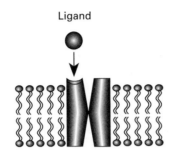

Ligand

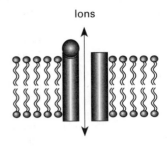

Ions

(b)

Ligand

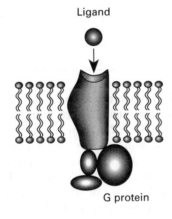

G protein

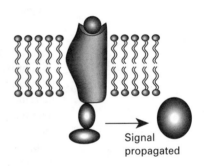

Signal propagated

(c)

Ligand

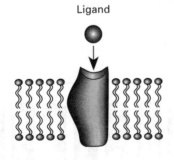

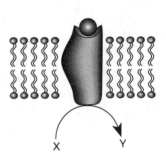

X   Y

Table 3.1. A few of the ligands that act through G protein-linked receptors.

Acetylcholine
Bradykinin
Calcitonin
Dopamine
Epinephrine (adrenaline)
Glucagon
Histamine
Leukotrienes
Lutropin
Neurotensin
Oxytocin
Parathyroid hormone
Prostaglandins
Retinal
Serotonin
Somatostatin
Thrombin
Thyrotropin
Vasopressin

In general, the topology of these receptors is such that they contain seven regions of approximately 22–24 amino acids, which form hydrophobic α-helices spanning the plasma membrane (Figure 3.2). Therefore their structure is analogous to that described for bacteriorhodopsin and they are commonly referred to as seven spanning receptors. The N-terminal end of the polypeptide is on the exterior face of the membrane but the C-terminal end is on the inside, which means that there are four cytoplasmic loop regions, the third of which is probably important for G protein binding, along with the cytoplasmic C-terminal tail. The class of G protein involved here is known as the trimeric G proteins, or heterotrimeric G proteins, which as their name suggests are composed of three subunits, α, β and γ. The functioning of these G proteins is further discussed in Chapter 5.

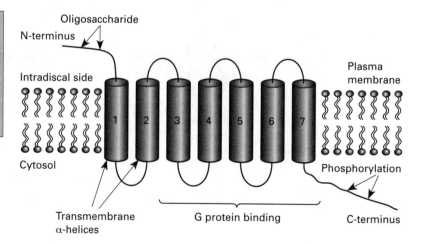

Figure 3.2
Predicted structure of a G protein-linked receptor, showing the seven α-helices spanning the membrane and the likely G protien-binding region.

As well as being activated by binding of the relevant ligand, the activity of these receptors is also altered by phosphorylation. Phosphorylation can be catalysed by cAMP-dependent protein kinase (cAPK) or by a class of kinases known as G protein-coupled receptor kinases (GRKs). GRKs are known to phosphorylate these receptors on multiple sites, using threonine and serine residues as targets. Phosphorylation deactivates the receptor as well as allowing the interaction of the receptor with an inhibitory protein known as β-arrestin. Therefore, such mechanisms are involved in the termination of the signal and receptor desensitisation. As discussed in Chapter 10, one of the classical examples of such a system is seen with rhodopsin, used in light perception in the eye. One of the most studied examples of these kinases is β-adrenergic receptor kinase (βARK); these kinases and their actions are further discussed in Chapter 4.

Ion Channel-linked Receptors

These receptors are often involved in the detection of neurotransmitter molecules where there is rapid signalling between neuronal cells excited by the passage of an electrical action potential, for example the acetylcholine receptor found at the nerve–muscle interface. Often these receptors are referred to as transmitter-gated ion channels. Binding of the ligand to the receptor changes the ion permeability of the plasma membrane as the receptor undergoes a conformational change that opens or closes an ion channel, but this is only a transient event, the receptor returning to its original state very rapidly. These receptors make up a family of related polypeptides, one of the distinguishing features being that they contain several polypeptide chains which pass through the membrane. For example, the acetylcholine receptor is composed of two identical polypeptides, which contain acetylcholine-binding sites, along with three different peptides, giving an α, α, β, λ, δ subunit structure (Figure 3.3). These five polypeptides are therefore encoded by four separate genes, which show a large degree of homology suggesting that they probably arose from gene duplication. In the membrane the proteins are arranged in a ring, in a similar fashion to that seen with gap junctions, and therefore there is a water-filled channel running through the middle from one side of the membrane to the other that allows the passage of ions. The ions passing through the hole made by the acetylcholine receptor are usually positively charged, such as Na^+, K^+ or Ca^{2+}, as the presence of negatively charged amino acids at the ends of the hole bestow some selectivity to the channel.

Receptors in this class, which possess different ligand-binding specificities and can be divided into subsets, all contain homologous polypeptides that arise from homologous genes or from alternative splicing. Therefore a vast array of receptors can be made using combinations of these different gene products. Often the presence of a particular receptor subset is tissue specific. It is these ion channel-linked receptors that are the targets of many drugs, such as barbiturates used in the treatment of insomnia, depression and anxiety.

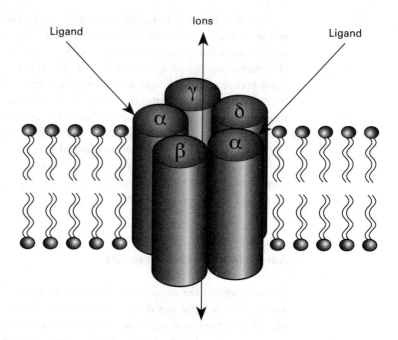

Figure 3.3
Predicted structure of an ion channel-linked receptor.

Receptors Containing Intrinsic Enzymatic Activity

These receptors comprise a quite heterogeneous class of receptors, characterised by the presence of a catalytic activity integral within the receptor polypeptide that is controlled by the ligand-binding event. The ligand-binding domain is found on the extracellular side of the membrane, with usually a single span of the membrane leading to the catalytic domain on the cytoplasmic side. The catalytic activity may be a guanylyl cyclase (discussed further in Chapter 5), a phosphatase or a kinase. The receptors may contain serine/threonine kinase activity and are often referred to as receptor serine/threonine kinases, or they may contain tyrosine kinase activity and are referred to as receptor tyrosine kinases (RTKs).

RTKs represent a class of these receptors that has been extensively studied. Ligand binding leads to activation of the kinase activity of the receptor, which causes phosphorylation on tyrosine residues of the receptor itself. This leads to the creation of binding sites particularly for proteins that contain protein-binding domains, for example SH2 domains. Such proteins may be adaptor proteins, such as GRB2 of mammals or Drk in *Drosophila*, which contain both SH2 and SH3 domains. On binding of such adaptor proteins to the receptor, complexes are formed that may also include a GNRP, for example Sos. Here, the result would be the activation of a G protein. This in turn may lead to a transduction cascade that could include further kinases, such as MAP kinases. However, phosphorylation is not confined to autophosphorylation and other proteins are also phosphorylated, leading to further propagation of the signal, for example the IRS-1 protein is phosphorylated on multiple sites by the insulin RTK as discussed in Chapter 9.

Over 50 RTKs have been identified and these can be grouped into at least 14 different families. Most classes of RTK are monomeric in nature, but some share the insulin receptor's tetrameric topology of an $\alpha_2\beta_2$ structure held together by disulphide bonds. Some RTKs contain cysteine-rich extracellular domains, while others contain extracellular antibody-like domains. However, they all seem to share the characteristics of having their N-terminal ends on the extracellular side of the membrane and the polypeptides only cross the membrane once. The signalling of these receptors is further discussed in Chapter 4.

These receptors are often found as dimers, ligand binding leading to the dimerisation event. However, some receptors are constitutively dimers, as seen with the insulin receptor: the receptor is coded for by one gene leading to the formation of one mRNA, but the protein product undergoes post-translational modification, including a cleavage event, and the two subsequent polypeptide chains are held together by the formation of cystine or disulphide bridges. However, dimerisation is a common theme in the activation of these receptors: the dimer may be a homodimer, i.e. containing two identical receptor subunits, or a heterodimer where the complex is composed of two different subunits from the same receptor family or even an accessory protein, for example the protein gp130. The formation of heterodimers allows the creation of a wide diversity of receptor specificities, as seen with PDGF receptors where the receptor dimers can be $\alpha\alpha$, $\beta\beta$ or $\alpha\beta$, each having different specificities to the isoforms of the growth factor.

Receptors Linked to Separate Tyrosine Kinases

Several receptors do not themselves contain a tyrosine kinase domain but on activation by ligand binding they cause the stimulation of a tyrosine kinase normally resident in the cytoplasm of the cell. Receptors for many growth factors and cytokines fall into this class, which is commonly referred to as the cytokine receptor superfamily. Again, the binding of the ligand to the receptor induces dimerisation. IFN-γ or human growth hormone cause homodimerisation of such receptors, while other ligand binding causes heterodimerisation or dimerisation involving the accessory protein gp130. However, the receptor for TNF-β forms trimers, while other ligands can lead to the formation of heterooligomers of three different polypeptides.

Activation of the receptor and complex formation often leads to the recruitment and stimulation of a family of soluble protein tyrosine kinases called JAKs that leads to the propagation of the signal (discussed further in Chapter 4).

Steroid Receptors: Intracellular Receptors of Extracellular Signals

Not all extracellular signalling molecules are detected on the surface of the cell by plasma membrane-borne receptors. The receptors for steroid hormones, thyroid hormones, retinoids and vitamin D are intracellular and commonly referred to as the intracellular receptor superfamily or steroid hormone receptor superfamily. These signalling molecules are all small and hydrophobic, and can readily diffuse across the plasma membrane. Steroid hormones are derived from cholesterol and include cortisol, vitamin D and steroid sex hormones. The amino acid tyrosine is the base for the thyroid hormones, while the retinoids are derived from vitamin A.

As these molecules are readily soluble in the hydrophobic environment of the membrane allowing their free passage into the cell, they are therefore inherently insoluble in the aqueous fluids outside the cells, such as the bloodstream. However, their solubility in water is increased by their association with specific carrier proteins. Dissociation from the carrier occurs before the signalling molecules enter the cell.

The receptors are found in either the cytosol or the nucleus of the cell. The nuclear ones are often found associated with DNA, albeit in an inactive state or acting in a gene-silencing role. The receptors can therefore be divided into two loose groups. The first group is found normally as an inactive complex involving the heat-shock proteins, Hsp90, Hsp70 and Hsp56. However, on binding to the ligand the receptors undergo conformational change and dissociation from the inhibitory Hsp complex, allowing their binding to specific sequences of bases on nuclear DNA, the so-called hormone response elements (HREs), and so altering the rates of transcription of specific genes (Figure 3.4). The receptors bind to DNA either as homodimers or heterodimers. The second group includes the receptors already found in the nucleus and associated with DNA. They have no Hsp associated with them.

Transcription rates may in fact be increased or decreased, depending on the receptor and gene sequence concerned. In some cases the cellular response to the signal can be twofold. A primary response is seen that manifests itself as an increase in the transcriptional rates of a small number of specific genes. The polypeptides produced by this first wave of accelerated transcription can themselves then cause later activation or reduction of the transcriptional rates of a further set of genes. Other gene regulatory proteins may also be involved and therefore the cellular response to these intracellular hormone receptors can be very tissue specific.

The steroid hormone receptor superfamily represents the largest known family of transcription factors described for eukaryotes. These include putative receptors identified through the use of sequence homology for which ligand specificity has not yet been determined, so-called orphan receptors, as well as numerous isoforms of known receptors. In general, amino acid sequences show that the receptors can be divided into several domains. At the N-terminal end of the polypeptide is a variable region known as the A/B domain, which is involved in gene activation. The next domain, towards the C-terminal end of the polypeptide, the C domain, contains two zinc fingers and is responsible for DNA recognition and for dimerisation. The next domain, the D domain, consists of a variable hinge region and may contain sequences responsible for localisation of the receptor to the nucleus. Ligand binding is to a large E domain. This region is approximately 250 amino acids in length and is also responsible for association with heat shock proteins and dimerisation. At the C-terminal end is the last region called the F domain, but no function has been assigned to this area of the protein.

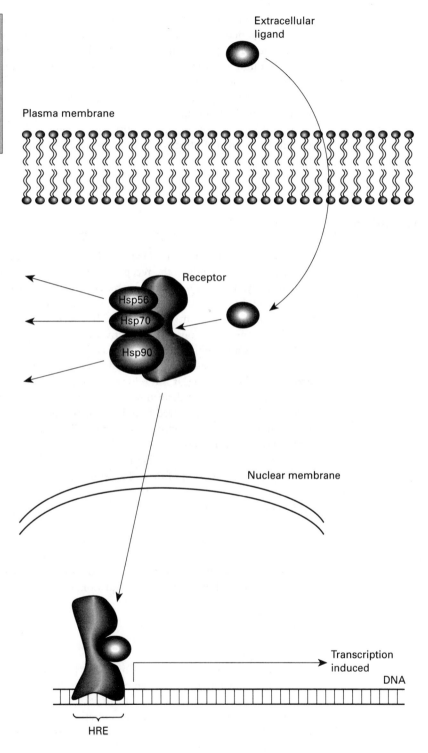

Figure 3.4
Schematic representation of the detection of lipophilic hormone molecules by an intracellular receptor. HRE, hormone response element; Hsp, heat-shock protein.

The region of DNA recognised by the receptors, the HRE, has been determined for the glucocorticoids to be two short imperfect inverted repeats with three nucleotides separating them. Other HREs have been found to be similar to this pattern, although it is the exact nucleotide sequences involved that bestow specificity on the interaction.

Steroid receptors also undergo a large amount of phosphorylation, both in a ligand-dependent and ligand-independent manner and it has been suggested that phosphorylation may not be essential for receptor function but might serve to enhance the normal effect or may be involved in the localisation of the receptor within the cell.

Ligand Binding

Ligands that bind to receptors can be classified under general headings. A ligand termed an agonist binds to the receptor and results in the activation of that receptor. Alternatively, a ligand referred to as an antagonist binds the receptor but does not result in activation of the receptor; further, the presence of an antagonist may interfere and stop the action of an agonist. If a receptor is active in the absence of a ligand, it is said to be constitutively active, a situation seen with some oncogenes.

Receptors are highly specific and usually have a high affinity for their respective ligand. However, the binding site of the receptor can be viewed as being like the active site of an enzyme. It is the local environment of the amino acids which make up the binding site that determines both the specificity and affinity for the ligand. The individual forces involved in holding the ligand on to the receptor are generally weak, being ionic attractions, van der Waals' forces, hydrogen bonding or hydrophobic interactions.

Ligand binding is a reversible reaction, allowing the receptor to be used and reused over a long period of time. Therefore the binding reaction can be written as:

$$L + R \rightleftharpoons LR$$

Just as with enzyme kinetics, one of the calculations useful in the determination of ligand-binding characteristics is the concentration of ligand at which half the receptors are bound, with half the receptors in the unbound state. This value is called K_d and can be defined by the following equation:

$$K_d = \frac{[R]\,[L]}{[RL]}$$

where [R] is the concentration of receptor, [L] is the concentration of ligand and [RL] is the concentration of receptor bound to ligand as a complex. The lower the K_d value, the higher the affinity of the receptor for its ligand. Usually K_d values approximate to the physiological concentrations of ligand, allowing the receptor to have the highest sensitivity to changes in the ligand concentration in the concentration range usually found.

Ligand binding is usually studied using a ligand that has been labelled or tagged, so that binding to its receptor can be followed. To visualise ligand binding to a cell-surface receptor for example, a fluorescent label may be used in conjunction with a fluorescence microscope or a confocal microscope. However, to quantify ligand binding a radiolabel is usually employed. The most common radiolabel is either ^{131}I or ^{125}I. However, these isotopes have extremely short half-lives, approximately 8 days and 60 days respectively. Further, the presence of a large iodine molecule may well interfere with ligand–receptor interaction. ^{3}H is also used as a label in many instances, one advantage being its long half-life of about 12 years.

If we consider a cell-surface receptor, a usual experiment would quantify the amount of ligand bound as the concentration of the ligand was increased. This would result in the total binding curve depicted in Figure 3.5. Binding measure-

Figure 3.5
A plot of the amount of hormone bound vs the hormone concentration, showing the total binding and the proportions made up by specific binding and non-specific binding.

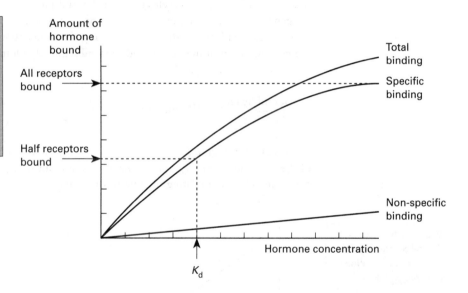

ments under the conditions usually adopted means that binding has come to equilibrium, but it should be noted that the speed of binding may be affected by the pH of the solution or the temperature of the reaction. Furthermore, some of the larger ligands might well have binding that is considerably slower and therefore care needs to be exercised if a true measure of binding is to be obtained. Once ligand binding has been quantified, the total amount of ligand bound does not mean that all the ligand has bound to the receptor, but includes an element of non-specifically bound ligand. The contribution of non-specific binding to the total is usually estimated by the inclusion of control experiments in which a huge excess of unlabelled ligand, approximately 100-fold excess, is added. This means that high-affinity receptor sites will be saturated with unlabelled ligand and therefore any labelled ligand remaining bound will be due to binding to non-specific sites. Non-specific binding is usually linear in a concentration-dependent manner. Once these values are subtracted from the total binding curve, specific binding to the receptor can be seen (Figure 3.5). This starts off being very much greater than non-specific binding, due to the high affinity of the receptors, but like typical enzyme kinetics the receptors become saturated and the binding curve tails off to a maximum, beyond which no more ligand binding occurs despite the addition of more ligand. Therefore, total specific binding sites and the K_d value can be estimated from the graph.

Estimating binding characteristics from a curve can be difficult and not very meaningful; therefore, in an analogous way to the analysis of enzyme activity, the ligand-binding data needs to be mathematically manipulated to give a linear relationship. Such analysis should give a better insight into the characteristics of the binding, for example whether the binding shows any cooperativity. The two common methods employed are derived from those developed by Hill and Scatchard. The original development of the Hill plot was to analyse the binding of oxygen to haemoglobin, but the same rationale can be used here.

The Scatchard plot is drawn by dividing the concentration of bound ligand by the concentration of free ligand and then subsequently plotting this against the concentration of bound ligand, i.e.:

$$\frac{[\text{Bound ligand}]}{[\text{Free ligand}]} \text{ vs } [\text{Bound ligand}]$$

As illustrated in Figure 3.6, from this plot the K_d value can be obtained, as the slope of the line will be $-1/K_d$ and the maximum amount of ligand bound (B_{max}) can be determined by extrapolating the line to the x-axis.

Figure 3.6
Typical Scatchard plot showing extrapolation of the slope to the x-axis.

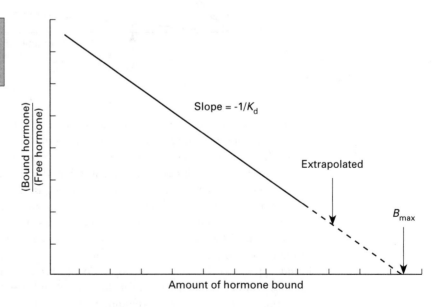

Although in theory the Scatchard analysis should give a straight line, this is quite often not the case. Disruption of ligand interaction with the receptor, for example by interference of the label used on the ligand, may result in curvature of the line.

In addition to artefactual reasons, the Scatchard analysis may not be linear for other reasons. Binding of the ligand to the receptor may show negative co-operativity, where binding of the first ligand to the receptor causes the receptor to have reduced affinity for the second ligand. Alternatively, there may be more than one type of receptor present, each with different binding characteristics, or the interaction of an intracellular subunit with the receptor, such as a G protein, reduces the affinity for the ligand. If Scatchard analysis is done when any of these factors is involved the resultant line will be curved and lie under the expected straight line (Figure 3.7b). Alternatively the line obtained may be curved but lie above the expected linear plot (Figure 3.7a). This may be an indicator of positive cooperativity, where binding of the first ligand increases the receptor's affinity for a second ligand, very much the same as the binding of oxygen to haemoglobin. Such effects may be due to conformational changes within the protein on ligand binding and may involve more than one polypeptide, i.e. the interaction of receptor molecules that would be possible with receptor dimerisation and complex formation.

Receptor Sensitivity and Receptor Density

The concentration or density of the receptors on the surface of a cell is not necessarily constant and it is very apparent that a cell may become more sensitive or less sensitive to a given extracellular concentration of ligand.

An increase in ligand sensitivity, or sensitisation, occurs by an increase in the amount of a receptor on the cell surface. Here a cell is maximising its chance of detecting the ligand and so responding to it. A real increase in the receptor available can be achieved by the synthesis of new receptor molecules, their recruitment from an intracellular store, such as from vesicles, or decrease in the rate of removal of the receptor from the cell surface, with one or more of these methods being responsible.

In many cases, if a cell has been exposed to a given ligand with a given response and shortly afterwards receives a second exposure to the same ligand, the response seen is very much reduced. The cells are said to become refractory to the second dose of ligand (Figure 3.8). If the cell retains a normal response to other ligands, i.e. only one receptor seems to be involved in the refractory state, the phenomenon is called homologous desensitisation. An example of this is seen with the β_2-adrenergic receptor and its response to ephinephrine. The receptor becomes desensitised extremely rapidly and it has been shown that both Mg^{2+} and ATP are required, suggesting the involvement of a phosphorylation step. It is now clear that there are in fact two separate phosphorylation events. Firstly, the β_2-adrenergic receptor causes a rise in cAMP and the activation of cAPK, which phosphorylates the receptor on a serine residue and disrupts the activation of the G protein by the receptor. Secondly, once activated, the receptor becomes a target for a specific kinase, βARK, which phosphorylates the receptor on several threonine and serine residues towards the C-terminal end of the polypeptide. This phosphorylated polypeptide can then bind to a protein called β-arrestin, which stops the receptor activating the G protein and so

Figure 3.7
Typical Scatchard plots where ligand binding is not straightforward. A curve is obtained for both positive cooperative binding (a) and negative cooperative binding (b).

(a)

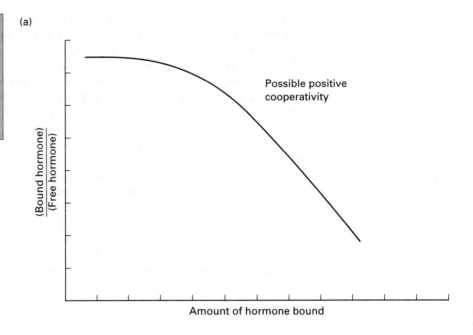

$\dfrac{\text{(Bound hormone)}}{\text{(Free hormone)}}$

Possible positive cooperativity

Amount of hormone bound

(b)

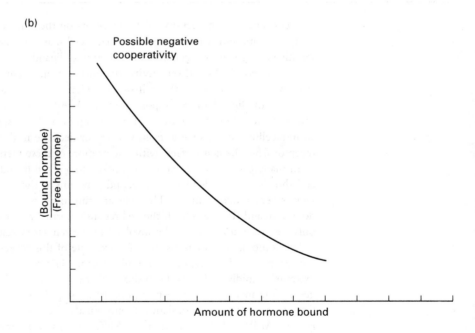

Possible negative cooperativity

$\dfrac{\text{(Bound hormone)}}{\text{(Free hormone)}}$

Amount of hormone bound

prevents the propagation of the response. It is interesting to note that rhodopsin, which shows structural similarity to the β-adrenergic receptor, also undergoes a similar desensitisation involving an arrestin-type mechanism.

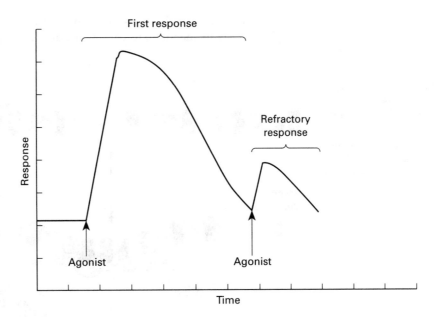

Figure 3.8
A plot of the hormonal response vs time showing that the application of a second dose of hormone does not necessarily invoke the same magnitude of response.

Since the β_2-adrenergic receptor can be phosphorylated in response to a rise in cAMP levels, which is controlled by many other receptors as well, it may well be desensitised in response to one of these other receptors binding to its respective ligand. Such a response is known as heterologous desensitisation, as it is not caused by a ligand normally associated with that receptor.

A common sequence of events on ligand binding may be as follows. Once a cell-surface receptor has been activated by the particular ligand and has transmitted its message to the next element of the signal-transduction cascade, for example via a G protein, the receptor is internalised into the cell by endocytosis. The membrane-bound receptor and ligand complex become an integral part of the vesicle that is formed and this is translocated and becomes part of an endosome. On the cell surface the receptor faces outwards, but in the formation of the vesicle the receptor ends up facing the inside of the endosome and is therefore exposed to the internal environment of the endosome, which is usually acidic. Therefore the receptor's affinity for the ligand is reduced and the receptor–ligand complex dissociates. For the receptor to be reused, it is once again transported by the vesicular system of the cell, this time back to the plasma membrane where it can once again be used to detect the extracellular presence of the ligand. The ligand left behind in the endosome is usually delivered to lysosomes where it is degraded. Hence the cells effectively remove the ligand from the extracellular fluids and cause the ligand signal to be turned off, unless the ligand continues to be released from its source. Not all receptor–ligand complexes dissociate to allow the recycling of the receptor back to the plasma membrane; in many cases both the receptor and ligand are transported to the lysosomes and destroyed. To maintain the receptor concentration on the cell surface therefore requires *de novo* protein synthesis. The insulin receptor is an example of this: cells exposed to continually high levels of insulin may be refractory to the insulin concentration due to the loss of receptors.

51

The process of internalisation of receptors is preceded by their relocalisation in the plane of the membrane into clusters (Figure 3.9), a process commonly called

Figure 3.9
The process of endocytosis. Receptors firstly move to the site of invagination and then a vacuole is formed, aided by the presence of clathrin molecules.

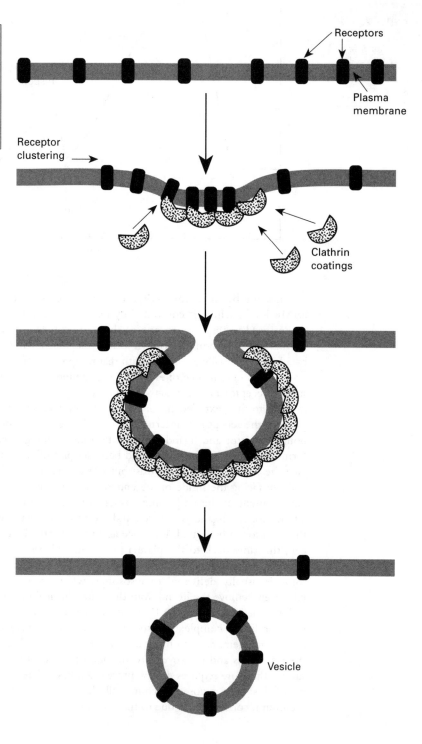

capping. This is followed by invagination of the membrane and the formation of what is known as a coated pit. This is so called because the cytoplasmic side of the forming vesicle is covered, or coated, with the protein clathrin. Clathrin is composed of two polypeptides, a light chain and a heavy chain; three of each come together to form a structure known as a triskelion. This is a three-legged structure that can further polymerise to form a basket-like matrix composed of pentamers and hexamers, which will form a scaffold around the vesicle (Figure 3.10). Clathrin

Figure 3.10
Likely quaternary structure of the clathrin protein, showing how the heavy chains and light chains align to give the appearence of a Manx flag, although the overall shape encloses a sphere.

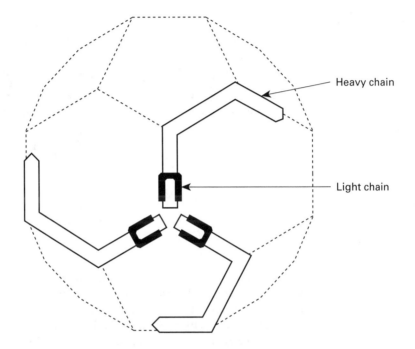

Heavy chain

Light chain

probably performs two main roles. Firstly, it propagates the formation of the invagination of the membrane and stabilises vesicle formation, as it naturally takes up a concave shape. Secondly, it may be involved in the capture of receptor molecules. Many receptors contain on their cytoplasmic side a short stretch of four amino acid residues, which acts as a signal for endocytosis. This short polypeptide region is recognised by a group of proteins known as the adaptins. These recognise the receptor and aid the binding of clathrin to the membrane. Different adaptins recognise different receptors and so coordinate the internalisation process. Once formed the vesicles shed the clathrin coat, as illustrated in Figure 3.9. This process probably involves ATP and Hsp70. The control of uncoating is also thought to involve the concentration of Ca^{2+} in the cell, where local rises in concentration may be encountered as the vesicle is transported deeper into the cell.

Summary

The detection of extracellular signals and the transmission of that signal into the cell is the responsibility of proteins known as receptors. As can be seen in Figure 3.11, such receptors are commonly found on the plasma membrane of the cell where they are ideally placed to be in contact with their extracellular ligand. However, steroid receptors are found inside the cell, with the ligand itself traversing the membrane.

Figure 3.11
A simplified overview of the role of receptors

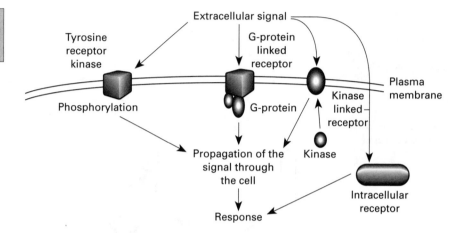

Membrane receptors fall into several groups: G protein-linked receptors, which lead to activation of the trimeric class of G proteins; ion channel-linked receptors, leading to changes in ion movements across the membrane; receptors containing intrinsic enzyme activity, such as the RTKs; and receptors that recruit separate enzymes, for example tyrosine kinases such as the JAK proteins.

Ligand binding to the receptor can be viewed as a reversible event, quantified by the K_d value, i.e. the concentration of ligand at which half the receptors are bound. However, the binding of the ligand to a cell usually includes two elements: specific binding to the receptor and non-specific binding.

Binding curves can be linearised by the use of Scatchard analysis, where the slope of the line is defined as $-1/K_d$. Such analysis can indicate if any cooperativity is involved in receptor–ligand binding.

A cell's response to the concentration of extracellular ligand can be varied by altering the density of receptors available at its surface. New receptor molecules can be synthesised and routed to the membrane or the response can be downregulated by capping and endocytosis, resulting in the internalisation of receptor molecules. Such internalised receptors can be returned to the membrane or alternatively destroyed.

Further Reading

Types of Receptors

Denis, M., Poellinger, L., Wikstrom, A. and Gustafsson, J. A. (1988) Requirement of hormone for thermal conversion of the glucocorticoid receptor to a DNA binding state. *Nature*, **333**, 686–8.

Evans, R. M. (1988) The steroid and thyroid hormone receptor superfamily. *Science*, **240**, 889–95.

Fath, I., Schweighoffer, F., Rey, I., *et al.* (1994) Cloning of a GRB2 isoform with apoptic properties. *Science*, **264**, 971–4.

Heldin, C.-H. (1995) Dimerization of cell surface receptors in signal transduction. *Cell*, **80**, 213–23.

Lee, C.-H., Kominos, D., Jaques, S., *et al.* (1994) Crystal structure of peptide complexes of the amino terminal SH2 domain of the SYP tyrosine phosphatase. *Structure*, **259**, 1157–61.

Lemmon, M. A. and Schlessinger, J. (1994) Regulation of signal transduction and signal diversity by receptor oligomerization. *Trends in Biochemical Sciences*, **19**, 459–63.

Orti, E., Bodwell, J. E. and Munck, A. (1992) Phosphorylation of steroid hormone receptors. *Endocrine Reviews.*, **13**, 105–28.

Ren, R., Mayer, B. J., Cicchetti, P. and Baltimore, D. (1993) Identification of a ten amino acid proline rich SH3 binding site. *Science*, **259**, 1157–61.

Sanchez, E. R. (1990) Hsp56: a novel heat shock protein associated with untransformed steroid receptor complexes. *Journal of Biological Chemistry*, **265**, 22067–70.

Strader, C. R., Fong, T. M., Tota, M. R. and Underwood, D. (1994) Structure and function of G protein-coupled receptors. *Annual Review of Biochemisty*, **63**, 101–32.

Strahle, U., Klock, G. and Schutz, G. (1987) A DNA sequence of 15 base pairs is sufficient to mediate both glucocorticoid and progesterone induction of gene expression. *Proceedings of the National Academy of Sciences USA*, **84**, 7871–5.

Tsai, M.-J. and O'Malley, B. W. (1994) Molecular mechanisms of action of steroid/thyroid receptor superfamily members. *Annual Review of Biochemistry*, **63**, 451–86.

Turner, A. J. (ed.) (1996) *Amino Acid Neurotransmission*. Portland Press, London.

Weigel, N. L., Carter, T. H., Schrader, W. T. and O'Malley, B. W. (1992) Chicken progesterone receptor is phosphorylated by a DNA dependent protein kinase during *in vitro* transcription assays. *Molecular Endocrinology*, **6**, 8–14.

Ligand Binding

Hill, A. V. (1913) The combination of haemoglobin with oxygen and carbon monoxide. *Biochemical Journal*, **7**, 471–80.

Scatchard, G. (1949) The attraction of protein for small molecules and ions. *Annals of the New York Academy of Sciences*, **51**, 660–72.

Receptor Sensitivity and Receptor Density

Anderson, R. (1992) Dissecting clathrin-coated pits. *Trends in Cell Biology*, **2**, 177–9.

Benovic, J. L., Mayor, F. and Somers, R. L. (1986) Light dependent phosphorylation of rhodopsin by β-adrenergic receptor kinase. *Nature*, **321**, 869–72.

Benovic, J. L., Mayor, F., Staniszeski, C., Lefkowitz, R. J. and Caron, M. G. (1987) Purification and characterisation of the β-adrenergic receptor kinase. *Journal of Biological Chemistry*, **262**, 9026–32.

Brodski, F. M., Hill, B. L., Acton, S. L., *et al.* (1991) Clathrin light chains: array of protein motifs that regulate coated-vesicle dynamics. *Trends in Biochemical Sciences*, **16**, 208–13.

Hausdorff, W. P., Caron, M. G. and Lefkowitz, R. J. (1990) Turning off the signal: desensitisation of β-adrenergic receptor function. *FASEB Journal*, **4**, 2881–9.

Keen, J. H. (1990) Clathrin and associated assembly and disassembly proteins. *Annual Review of Biochemistry*, **59**, 415–38.

Lefkowitz, R. J. (1993) G-protein-coupled receptor kinases. *Cell*, **74**, 409–12.

Nathke, I. S., Heuser, J., Lupas, A., Stock, J., Turck, C. W. and Brodsky, F. M. (1992) Folding and trimerisation of clathrin subunits at the triskelion hub. *Cell*, **68**, 899–910.

Palczewski, K. and Benovic, J. L. (1991) G-protein-coupled receptor kinases. *Trends in Biochemical Sciences*, **16**, 387–91.

Pearce, B. M. (1989) Characterisation of coated-vesicle adaptins: their assembly with clathrin and with recycling receptors. *Methods in Cell Biology*, **31**, 229–46.

4 Protein Phosphorylation, Kinases and Phosphatases

Topics

- Serine/Threonine Kinases
- Tyrosine Kinases
- Mitogen-Activated Protein Kinases
- Histidine Phosphorylation
- Phosphatases
- Other Covalent Modifications

Introduction

The arrival of an extracellular signal at the cell's surface and its detection by binding to a receptor must lead to further events inside the cell if the cell is to respond to the presence of that ligand. Activation of the receptor allows it to interact with and/or activate several different types of protein, which lead to a wide range of intracellular signalling cascades as discussed in subsequent chapters. However, the final event in nearly all cell-signalling pathways is the modification of the activity of an enzyme or activation factor.

An enzyme's specific activity may be altered by a change in its conformation, i.e., the folding of the polypeptide chain. This is usually seen as a change in its tertiary structure, although in multipolypeptide enzymes it may also involve the quaternary structure of the protein. The altered spatial arrangement of the active site amino acids will then reduce or increase substrate binding and/or the catalytic action of the protein.

One of the most common ways of modifying protein structure is the addition of one or more phosphate groups to the primary amino acid sequence of the polypeptide, a process known as phosphorylation. For phosphorylation to be an effective control mechanism allowing the activity of an enzyme to be both increased and decreased, the overall reaction has to be reversible (Figure 4.1). If, for example, the need for glycogen breakdown is suddenly increased due to a necessity to supply the muscles with glucose and hence energy, the enzyme responsible, phosphorylase, becomes phosphorylated with a concomitant increase in its activity. However, on cessation of the muscle's activity, the need for glucose would drop and therefore the rate of glycogen breakdown would have to be reduced. A quick and energetically favourable way to do this would be dephosphorylation of phosphorylase thus lowering its activity, rather than, for example, destruction of the phosphorylase polypeptide to remove its activity.

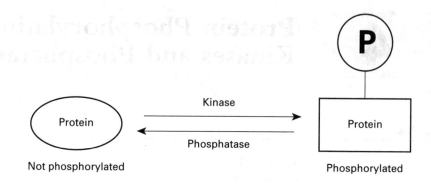

Figure 4.1
Phosphorylation and dephosphorylation of a protein.

The coordinated control of metabolic pathways also requires that futile cycles are avoided. Again using glycogen storage as an example, to facilitate glycogen usage phosphorylase is phosphorylated causing an increase in its activity, as already mentioned. However, there would appear to be little point in increasing glycogen breakdown if its synthesis continued unabated or in fact increased due to the rise in the concentration of intracellular glucose resulting from enhanced glycogen breakdown. Indeed, this is coordinated by phosphorylation of the enzyme responsible for the synthesis of glycogen, glycogen synthase, at the same time, but in this case phosphorylation causes a lowering of enzyme activity, slowing the production of glycogen. Therefore, by phosphorylation of the two key enzymes in the pathway at the same time breakdown is increased and synthesis decreased, resulting in the necessary effect (Figure 4.2). Some of the enzymes involved in metabolic pathways and which are controlled by phosphorylation and dephosphorylation are listed in Table 4.1.

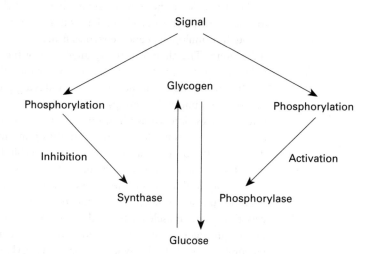

Figure 4.2
Phosphorylation controls both glycogen breakdown and synthesis.

Table 4.1. Some of the enzymes which are controlled by phosphorylation/dephosphorylation and the metabolic pathway in which they are involved.

Enzyme	Metabolic pathway
Phosphorylase kinase	Glycogen usage
Glycogen phosphorylase	Glycogen usage
Glycogen synthase	Glycogen storage
Acetyl-CoA carboxylase	Lipid metabolism
Lipase	Lipid breakdown
Isocitrate dehydrogenase	Citric acid cycle/in prokaryotes

The enzymes that catalyse protein phosphorylation are known as protein kinases and there is a reciprocal group of enzymes that carry out dephosphorylation, called phosphatases. As phosphorylation takes place on one of three amino acids in the primary sequence of the polypeptide (serine, threonine or tyrosine; Figure 4.3), the

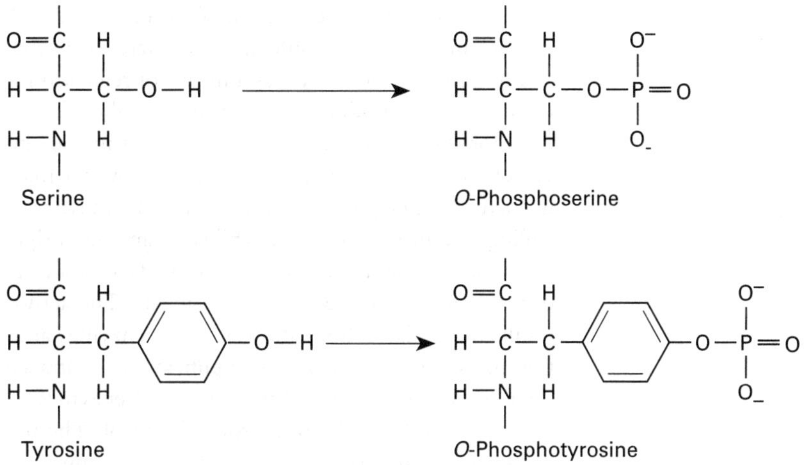

Figure 4.3
Phosphorylation of serine and tyrosine yields the altered residues phosphoserine and phosphotyrosine.

kinases are grouped according to which amino acid they are specific for, i.e. serine/threonine kinase or tyrosine kinase. The phosphate group is supplied by ATP, the third phosphoryl group (γ) of the chain being transferred to the hydroxyl group of the acceptor amino acid, with the subsequent release of adenosine diphosphate (ADP). Phosphorylation takes place inside the cell where ATP is generally abundant. Dephosphorylation is the simple removal of the phosphoryl group from the amino acid, with the regeneration of the hydroxyl side-chain and the release of orthophosphate. Each enzyme catalyses their relevant reaction in an irreversible manner, although the overall reaction of phosphorylation and the return of the protein to its original state is effectively accompanied by the hydrolysis of ATP and hence results in a favourable free-energy change. Phosphorylation of the protein is not restricted to a single site on the polypeptide chain and indeed a protein may be phosphorylated by more than one kinase, allowing in many cases the convergence of several signalling pathways. Glycogen synthase can be phosphorylated by cAPK, phosphorylase kinase and Ca^{2+}-dependent kinase. Each phosphorylation event might have a different effect on the protein or may have no effect at all. Phosphorylation of serine 10 on cAPK and on the protein designated as p47-*phox* of the NADPH oxidase enzyme system, which seems to occur on activation of the system, appears to serve no purpose as phosphorylation bestows no alteration of activity on these polypeptides.

The phosphorylation of a polypeptide can alter the enzymatic activity of that molecule and this might be achieved for several reasons. The added phosphoryl group supplies negative charge to an enzyme, which can disrupt electrostatic interactions or may be instrumental in the formation of new interactions. Similarly, the phosphoryl group can form hydrogen bonds, which may favour a new conformation. The free-energy change involved in phosphorylation may also help to push the equilibrium from one conformational state to another.

Phosphorylation is ideal as a means of regulation in response to a cellular signal as it can occur in under 1 s, as can dephosphorylation, therefore making the system ideal for the fast interaction often needed in the alteration of metabolic rate. Conversely, the process may have kinetics over a matter of hours, which might be needed in other physiological conditions. Further, one of the basic needs of a signalling system in a cell is the ability to amplify a signal. As discussed in Chapter 1, a few molecules arriving on the outer surface of a cell might need to alter the activity of many enzyme molecules on the inside. The activation of a single kinase molecule results in the phosphorylation of many enzymes and so the process of phosphorylation plays a major role in the amplification of intracellular signals.

If kinases are to control the activity of enzymes within the cells, then they themselves have to be under some sort of control. The modulation of a cell's activity may be through various different routes, including alteration of Ca^{2+} concentrations, cAMP, cGMP and inositol phosphate metabolism. These different signalling pathways commonly appear to culminate with the activity of a particular type of kinase, with their own degree of specificity towards the enzymes that they control. In fact, some kinases might themselves be controlled by a phosphorylation event, for example phosphorylase kinase is itself phosphorylated by cAPK. However, not all kinases are controlled by second messengers, a classic case of messenger-independent protein kinases being the casein kinases. These enzymes are widely distributed throughout the plant and animal kingdoms, where they are used for the phosphorylation of acidic proteins.

Some kinases are specific for one protein, for example phosphorylase kinase, while others have a more generic action, being capable of phosphorylating many proteins, for example PKC. The specificity of the kinase is determined by the specific amino acid sequence either side of the target amino acid residue destined to receive the phosphoryl group. In many cases, consensus sequences for the sites of phosphorylation have been identified, but the appearance of a consensus sequence in the protein's primary sequence does not automatically mean that the site will be phosphorylated. The overall conformation of the protein might be such that the target sequence has been masked and rendered unusable.

Although kinases have been separated into these two broad classes, the actual catalytic site seems to be quite well conserved amongst these enzymes. There appear to be several conserved regions in the catalytic region but two particular regions have had conserved sequences or signature patterns assigned for them. The first region is located at the N-terminal extremity of the catalytic region. It is characterised by a lysine residue believed to be involved in ATP binding that is close to a stretch of glycine residues. In the central part of the catalytic region the second conserved region to be identified contains an aspartic acid residue believed to be important in the catalytic activity of the kinase. However, the amino acids around this residue appear to differ between the two classes of kinase and two signature patterns have been deduced. However, not all kinases contain these regions, although interestingly the signature pattern specific for the tyrosine kinase catalytic site has homology to some bacterial phosphotransferases, which are thought to be evolutionarily related and contain some structural homology to protein kinases. The identification of such protein signatures will aid in the identification of kinase active sites in newly discovered proteins. It has in fact been estimated that the human genome may contain as many as 2000 genes coding for different kinase polypeptides, leaving many to be discovered.

Serine/Threonine Kinases

The kinases that perferentially phosphorylate serine or threonine residues and therefore come under the classification of serine/threonine kinases encompass a large group of phosphorylating enzymes, including cAPK, cGPK, PKC, $Ca^{2+}/$ calmodulin-dependent protein kinases, phosphorylase kinase, pyruvate dehydrogenase kinase and many others. Not all these can be treated in detail here, but important representative examples are examined that should give an overall picture of the activity of these ubiquitous enzymes.

Several oncogenes appear to function because they encode proteins containing serine/threonine kinase activity. These include the products of the genes *mil*, *raf* and *mos*. Such activity highlights the importance of phosphorylation by these kinases in the control of cellular functions.

cAMP-dependent protein kinase

cAPK (or protein kinase A, PKA) is widespread in eukaryotes, being found in animals, fungi and as a slightly different form in plants. It is not found in prokaryotes.

Many processes within the cell are controlled through the activation of cAPK. These include the regulation of metabolism and control of gene expression. Examples of such control include activation of phosphorylase kinase and hence activation of phosphorylase. This results in the increased breakdown of glycogen, which is concomitant with the phosphorylation and deactivation of glycogen synthase (see Figure 5.1). Gene expression can be controlled through the phosphorylation and activation of transcription factors such as CREB, which binds to the cAMP-response element (CRE) regions of the DNA. Genes containing such control elements include those that encode enzymes of gluconeogenesis in the liver.

In the inactive state, i.e. in the absence of cAMP, cAPK is found as a tetramer of two catalytic subunits (C) and two regulatory subunits (R). On activation, cAMP binds to the regulatory subunits causing an alteration in the affinity of the regulatory subunits for the catalytic subunits, causing the complex to dissociate. The regulatory subunits remain a dimer with the release of the two active monomeric catalytic sub-units (Figure 4.4). It is thought that virtually all of the cell's responses to cAMP are

Figure 4.4
Activation of cAMP–dependent protein kinase, showing the dissociation caused by the binding of cAMP. C, catalytic subunit; R regulatory subunit.

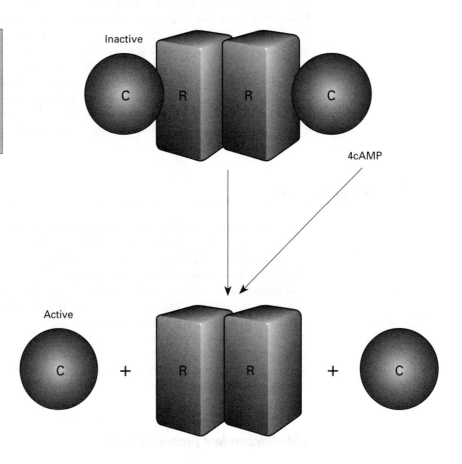

mediated by the activity of the catalytic subunits of cAPK. As a monomer, this polypeptide has a molecular mass of around 41 kDa but at least three isoenzyme forms have been identified. The C_α and C_β subunits in mammals differ by less than

10% when their amino acid sequences are compared and they seem to be highly conserved between species. Unlike C_α, which appears to be expressed constitutively in most cells, C_β is tissue specific. The catalytic core of the subunits also shares homology with other known kinases and includes defined regions for peptide binding, ATP binding and a catalytic site.

Studies of the catalytic regions of the polypeptide, using fluorescence analogues of ATP and by studying the modification of lysine residues by acetic anhydride in the presence and absence of MgATP, have shown that lysine residues at amino acid positions 47, 72 and 76 are important. Sequence comparisons with other kinases showed that lysine 72 was totally invariant and site-directed mutagenesis has shown its vital importance in the catalytic cycle of kinases. To the N-terminal side of this lysine lie three highly conserved glycine residues at positions 50, 52 and 55, which are probably involved in binding of the phosphate in the nucleotide. Similarly, point mutations in an analogous glycine-rich region may be responsible for the constitutive activity of the v-*erbB* oncogene product, which resembles the EGF receptor but is in a permanently activated state. However, the v-*erbB* oncogene protein also lacks the extracellular EGF-binding domain as well.

cAPK typically phosphorylates peptides that contain two consecutive basic residues, usually arginine, which lie at positions 2 and 3 towards the N-terminal end of the protein from the site of phosphorylation. Thus the consensus sequences are:

-Arg-Arg-X-Ser-X-
-Arg-Arg-X-Thr-X-   or

However, there appear to be many exceptions to this rule. The residue designated as X between the Arg doublet and the phosphoylation site is usually a small amino acid, while the other amino acid designated as X is usually hydrophobic in character. The arginines may also be replaced by lysines. Futhermore, carboxyl groups within the catalytic subunit may well be responsible for this specificity, allowing the recognition of the peptide substrate. Carboxyl groups, particularly at positions 184 (aspartate), and 91 (glutamate), may also be responsible for ligation to the Mg^{2+} in the MgATP substrate. Like lysine 72, these two groups appear to be invariant throughout the kinase domains studied so far. Further carboxyl groups at position 170 and six at the C-terminal end of the protein have also been implicated in the recognition of the peptide substrate. Kinetic analysis of cAPK shows that it usually has a K_m in the region of 10–20 µmol/l substrate with a V_{max} of 8–20 µmol/min per mg protein.

The first step in binding of the peptide to the enzyme probably involves ionic interactions between the peptide's arginine residues and the enzyme, followed by recognition of the serine that will ultimately accept the phosphate group. Binding of MgATP to the catalytic subunit enhances binding of the peptide and large conformational changes across the enzyme are induced by binding of the substrate.

It is known that the catalytic subunit of cAPK contains two phosphorylation sites, one at threonine 197 and another at serine 338. Once phosphorylated the phosphate groups are not readily removed by phosphatases. A serine residue at position 10 can also be autophosphorylated, although it is not known if this has any physiological role. The catalytic subunit also is covalently modified by myristoylation of the N-terminus. It was thought that this might serve as a signal for translocation of the subunit to the plasma membrane, but the catalytic subunit remains soluble and myristoylation is not essential for catalytic activity, so the role of this modification is obscure.

The regulatory subunit of cAPK has the function of binding to the catalytic subunits so rendering them inactive. Two main groups of the regulatory subunits have been found, called type I and type II. They differ not only in their amino acid sequences but also in their function. Type II subunits can be autophosphorylated while type I are not autophosphorylated but contain a binding site for MgATP, which binds with relatively high affinity. Isoforms of the different types of subunit have been cloned, some being fairly ubiquitous while others show tissue specificity. Furthermore, some isoforms are inducible while others appear to be expressed constitutively.

In general, the regulatory subunit is a dimer, with all the subunits sharing the same structural features (Figure 4.5). These include two consecutive gene duplicated

Figure 4.5
Domain structure of the regulatory subunit of cAMP-dependent protein kinase.

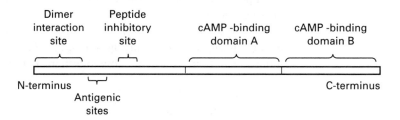

sequences at the C-terminal end of the molecule that are the cAMP-binding sites. Type I subunits are held together covalently by two disulphide bonds with the two polypeptide chains running antiparallel. In type II, the subunits are held together by interactions between the N-terminal amino acids in the chains. Another general feature of the subunits is what is referred to as the hinge region towards the N-terminal end, which is sensitive to proteolytic cleavage. This region encompasses the amino acids most likely to be involved in the interaction with other proteins, as well as being a highly antigenic part of the molecule.

The most important part of the regulatory subunit is a region that controls the activity of the catalytic subunit. The hinge region contains an amino acid sequence, known as the peptide inhibitory site, which resembles that of the substrate of the catalytic subunit. In type II regulatory subunits there is an autophosphorylation site here, while type I subunits have a sequence that contains the two arginines needed for recognition by the catalytic subunit active site but lack the serine or threonine which would normally accept the phosphate group. This so-called pseudophosphorylation site contains an inert alanine or glycine residue instead. Several lines of evidence, including limited proteolytic cleavage, affinity studies of the regulatory subunit to the catalytic subunit and site-directed mutagenesis, support the idea that the regulatory subunit maintains the catalytic subunit in an inactive state, whereby the pseudophosphorylation or inhibitor site of the regulatory subunit occupies the peptide-binding site of the catalytic subunit so preventing the binding of the correct protein substrate.

In this example, the inhibitory site and the catalytic site are on separate polypeptide chains, but this general principle of a regulator region controlling the activity of the catalytic region is used by several other protein kinases, although usually both sites are part of a single polypeptide chain. This includes the related kinase cGPK as well as others such as myosin light chain kinase (MLCK) and PKC. It is the binding of Ca^{2+}/calmodulin that causes a major conformational change in the MLCK molecule, preventing the inhibitor site from blocking the active site of the catalytic region. In other kinases the resemblance is more like cAPK type II regulatory subunits, in that the site is actually autophosphorylated; examples include cGPK and Ca^{2+}/calmodulin-dependent kinase II.

In cAPK the conformational change needed to release the inhibitor site from the catalytic active site is induced by the binding of cAMP. Each regulatory site has two cAMP-binding sites, both of high affinity. These two sites, A and B, show high sequence homology to each other but their binding to various cAMP analogues varies. However, they do show high sequence homology to catabolite gene activator protein (CAP) of *Escherichia coli*. This protein is involved in the cAMP-dependent regulation of the lactose operon, activating gene expression when the cells are depleted of glucose and allowing them to survive on a new carbon source. In CAP it was found that two amino acid residues were of crucial importance to the functioning of the protein: an arginine that interacts with the negative charge of the phosphate of cAMP and a glutamine residue that hydrogen bonds to the ribose ring. These two residues are found in the cAPK regulatory subunits as well as in the cGMP-binding sites of cGPK.

Deletion studies have shown that the N-terminal region of the regulatory subunit, along with the cAMP-binding site B, can be removed and the polypeptide is still able to bind to the catalytic subunit in a cAMP-dependent manner. Removal of the cAMP-binding site A has also been shown to produce a molecule that is cAMP dependent and can still bind to the catalytic subunit. Removal of either binding site seems to have little effect on the affinity for cAMP of the other binding site in the isolated regulatory subunits. However, it is thought that both cAMP-binding sites participate in the activation of the type I cAPK, with both sites needing to be occupied for dissociation of the native holoenzyme with binding of the two cAMP-binding sites being cooperative. cAMP probably binds to site B first, causing conformational changes that increase the accessibility of site A to cAMP. Binding here causes the conformational change in the molecule, altering the hinge region in particular. Several amino acids have been implicated to be involved here, but the exact mechanism is not known.

Reassociation of the cAPK complex for the type II regulatory subunit-containing enzymes probably involves the use of phosphatases, allowing the inhibitor site to once again enter the catalytic site and cause deactivation of the enzyme's activity. However the reassociation for type I regulatory subunit-containing enzymes involves the binding of MgATP. This binding shows positive cooperativity, which probably ensures that the holoenzyme is a tetramer and that trimers with only one catalytic subunit are not formed.

The N-terminal end of the regulatory subunits has also been seen to be the target for phosphorylation by several kinases, including casein kinase II and glycogen synthase kinase. Phosphorylation here might be involved in the association of the regulatory subunits with other proteins.

It has been found that some forms of cAPK are associated with the membrane fractions of cells and this is probably mediated by the regulatory subunits binding to an integral membrane protein. The catalytic subunits are probably not involved as they can be released and found in the soluble fractions of cells.

cGMP-dependent Protein Kinase

cAMP is not the only nucleotide responsible for the activation of a kinase. cGMP is also an important second messenger in cells, used to control phosphorylation via cGPK. This enzyme is abundant in smooth muscles, heart, lung and brain tissues. Although mainly cytosolic it has also been found in the particulate fraction after cell subfractionation. The cGPK purified from lung and heart is a dimer of identical 76-kDa subunits, with the holoenzyme having a molecular mass of approximately 155 kDa. The two subunits are held together in an antiparallel arrangement by disulphide bonds. In this orientation the inhibitor site of the regulatory domain of one subunit acts as the inhibitor of the catalytic domain of the other subunit, similar to the separate subunits of cAPK.

The cGPK sequence can be thought of as being six segments which make up four functional domains. The segment at the N-terminal end of the polypeptide contains the sites used to maintain the dimer structure, a hinge region as seen in cAPK, and the inhibitor site, which will undergo autophosphorylation. Also like cAPK, the next two segments have high sequence homology to the CAP protein of *E. coli* and comprise the two cGMP-binding domains of the polypeptide. The rest of the molecule is the catalytic domain. The fourth and fifth segments show high homology to other protein kinases.

Also like cAPK, cGPK shares very similar preferences for the peptide sequences that are phosphorylated. The enzyme typically phosphorylates a serine or threonine in peptides containing two consecutive basic residues, usually arginine, that lie at positions 2 and 3 towards the N-terminal end of the protein from the site of phosphorylation. Therefore, like cAPK, the consensus sequences are:

$$-Arg-Arg-X-Ser-X-$$
$$-Arg-Arg-X-Thr-X- \quad \text{or}$$

In cGPK, a proline residue also N-terminal to the phosphorylation site and a C-terminal basic residue enhance phosphorylation. Such differences give cGPK a different substrate specificity to cAPK.

Although activated primarily by cGMP, cGPK can also be activated by cAMP and cyclic inosine monophosphate (cIMP) at much higher concentrations. However, unlike cAPK, cGPK does not dissociate on activation, although like cAPK both nucleotide-binding sites need to be occupied for activation.

cGPK itself is also phosphorylated, but the physiological relevance of this is not known.

Protein Kinase C

PKC was originally thought to be just a single protein, but more recent research has revealed that it is in fact a family of closely related protein kinases. These different polypeptides are formed either as products of multiple genes or are derived from the alternative splicing of an mRNA transcript. However, different cells show variation in the expression of the isoenzymes, with some cells expressing more than one form.

PKC proteins can be broadly split into two families. The first to be cloned was the group containing the α, βI, βII and γ subspecies with more recent cloning revealing a family containing δ, ε, and ζ subspecies. The genes for the α, β and γ forms have been located on different chromosomes, with the variation in the βI and βII forms coming from alternative splicing. In general, PKC enzymes are monomeric in nature, having between 592 and 737 amino acids, giving molecular masses between 67 and 83 kDa. The two groups have a common structure and show close similarity to each other, although distinct differences are seen between the two families. The first group, containing the α, βI, βII and γ forms, have four conserved regions (C_1–C_4) along with five variable regions (V_1–V_5). The second group, containing the δ, ε, and ζ forms, lack the C_2 conserved region, although overall their molecular masses appear to be similar across the families. The polypeptides can be roughly divided into two domains. The regulatory domain contains the C_1–C_2, and V_1–V_2 regions of the α, βI, βII and γ group, but only the C_1 region of the δ, ε group (Figure 4.6). The C_1 region of both contains a highly cysteine-rich region

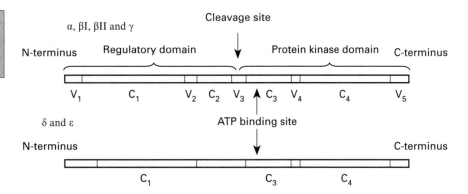

Figure 4.6
Domain structure of the two families of protein kinase C.

that resembles the consensus sequences of a cysteine–zinc DNA-binding finger, although no DNA binding has been reported. The second domain contains the rest of the molecule and is referred to as the protein kinase domain. Sequence homology between this region and other protein kinases has been reported while the C_3 region contains an ATP-binding sequence, a motif repeated although probably not used in the C_4 region. Recent studies using X-ray absorption also suggest that PKC may contain four zinc ions, which are coordinated mainly by sulphur atoms supplied by the cysteine residues. These probably serve a role in the structural stability of the polypeptide rather than having a direct role in the catalytic cycle.

Other isoforms described include η, θ and λ making a total of 10 isoforms, but there is no reason to suspect that more will not be discovered.

PKC can be activated by a wide range of stimuli, including by proteolytic cleavage or Ca^{2+} concentrations, which is how it derived its name as a Ca^{2+}-dependent protein kinase. However, some isoforms are Ca^{2+} independent, such as δ, η, θ and ε. PKC can also be controlled by phospholipids and in particular by DAG, which can be derived from inositol phosphate metabolism as discussed in Chapter 6. Chemical stimulants such as the tumour-promoting phorbol esters, for example phorbol 12-myristate 13-acetate (PMA) otherwise known as 12-O-tetradecanoyl-phorbol-12-acetate (TPA), are widely used in the laboratory to cause activation of PKC because they act as DAG analogues. However, their exact activation differs between isoenzyme forms and in fact some isoforms, such as ζ and λ, seem to lack the DAG-binding site. It has been suggested that some PKC subspecies become activated at different times during a cellular response, orchestrated by a series of phospholipid metabolites such as DAG, arachidonic acid or other unsaturated fatty acids. Caution is needed when using phorbol esters in experimental work, not only due to their toxicity to the user but also because they are only very slowly metabolised from the cells, unlike DAG. New stimulators more specific than phorbol esters, such as sapintoxin A, can also be used in the laboratory to activate PKC.

As in cAPK, it appears that PKC has an inhibitor site near the N-terminus of the polypeptide. Again, this site contains an inert alanine instead of the threonine or serine that would normally be found in the substrate. Binding of its activators, such as Ca^{2+}, DAG and phosphatidylserine, causes a conformational change releasing this pseudophosphorylation site and resulting in an active protein.

Activation by proteolytic cleavage can occur through the action of the enzyme calpain. Cleavage occurs within the variable V_3 region, releasing a catalytically active fragment. It may be that the active form of PKC is the target for calpain, which itself is active in micromolar Ca^{2+} concentrations. However, unlike the γ form, not all subspecies of PKC are susceptible to rapid cleavage, for example subspecies α is relatively resistant. It may be that cleavage of PKC is in fact not part of its activation but the first step in its degradation and removal from the cell.

Once activated, under physiological conditions, PKC preferentially phosphorylates a polypeptide on a serine or threonine residue found in close proximity to a C-terminal basic residue. The consensus sequence is thus:

-[Ser/Thr]-X-[Arg/Lys]-

Additional basic residues on either the C-terminal or N-terminal side of the target amino acid may enhance the V_{max} and K_m of the phosphorylation reaction.

The exact role of PKC within the cell remains relatively obscure. Phosphorylation experiments *in vitro* have shown that a large variety of polypeptides are phosphorylated by PKC and it has been implicated in activation of Ca^{2+}-ATPases and the Na^+/Ca^{2+} exchanger, which control Ca^{2+} levels within the cell, as well as the phosphorylation of receptors such as the EGF receptor and the IL-2 receptor.

Ca²⁺-calmodulin-dependent protein kinases

One of the major ways of controlling the function of enzymes and metabolic pathways in a cell is via alteration of the intracellular concentration of Ca^{2+}, $[Ca^{2+}]_i$. Many kinases have been found to be dependent on or regulated by Ca^{2+}, including PKC, MLCK and phosphorylase kinase. Besides these more specific enzymes there is a group of kinases that are controlled by Ca^{2+} called the multifunctional Ca^{2+}/calmodulin-dependent protein kinases. Several classes of these have been identified but the best characterised is known as Ca^{2+}/calmodulin-dependent protein kinase II. This enzyme is also referred to as CaM kinase II, type II CaM kinase or sometimes simply just kinase II.

Although kinase II is widespread throughout several tissues, it is most abundant, up to 20–50 times more concentrated, in brain and neuronal tissue compared with non-neuronal tissues. The enzyme prepared from brain tissue consists of an α subunit of approximately 50 kDa and a protein doublet (β/β′) of approximately 60 kDa. Both types of subunit are able to bind to calmodulin and are autophosphorylated in a Ca^{2+}/calmodulin-dependent manner. Both subunits therefore contain both regulatory and catalytic domains. The α and β/β′ subunits are very closely related, the N-terminal halves of the two subunits being 91% identical, with over three-quarters of the other half also being identical when the genes were cloned from rat brain. The holoenzyme has been reported to have a molecular mass of 500–700 kDa in most tissues that have been investigated but a smaller complex of 300 kDa has been found in liver. The complex probably assumes a dodecamer structure in which two hexameric rings are stacked on top of each other. The ratio of α subunits to β/β′ subunits seems to vary enormously between tissues while other subunits of approximately 20 kDa have been reported in spleen tissues. Isoenzymes cloned include δ and γ subunits which are of similar molecular mass to the β/β′ subunits and which appear to have a wide tissue distribution. The sequences of these isoforms show that they are also closely related to the α subunits.

The subunits can be split into three distinct functional domains. The N-terminal section is the catalytic domain, the centre section encompasses the regulatory domain and the C-terminal section is the domain essential for the formation of the holoenzyme complex, known as the association domain (Figure 4.7). Proteolytic

Figure 4.7
Domain structure of Ca^{2+}/calmodulin-dependent protein kinase II.

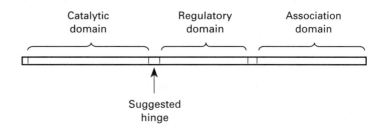

Catalytic domain   Regulatory domain   Association domain

Suggested hinge

cleavage can release an active catalytic domain and here homology to other calmodulin-regulated kinases is seen, such as phosphorylase kinase. The fact that proteases have activity here suggests the presence of a hinge region between the catalytic and regulatory domains. Like other kinases discussed, the regulatory domain contains an autoinhibitory site, but in this case a calmodulin-binding site is also found. Binding of Ca^{2+}/calmodulin, which shows positive cooperativity, causes an overall conformational change in the kinase's structure releasing the autoinhibitory site from the active site and allowing full kinase activity. In the inactive state the autoinhibitory site not only prevents peptide substrate binding but also prevents the binding of the donating ATP. The autoinhibitory site involves autophosphorylation as seen with other kinases and here, too, it is thought that this autophosphorylation potentiates the effects of the second messenger. It is thought that a rapid spike of Ca^{2+} activates the enzyme not only allowing it to phosphorylate its normal substrate but also causing autophosphorylation, which probably slows down calmodulin dissociation from the enzyme. This lengthens kinase activity even after Ca^{2+} concentrations have returned to their basal levels.

The subcellular location of the enzyme also seems to vary, sometimes being membrane associated, sometimes soluble or sometimes bound to the cytoskeletal structures of the cell.

The substrates for kinase II include tryptophan hydroxylase, glycogen synthase from skeletal muscle, synapsin I, proteins associated with microtubules, ion channels and transcription factors. Peptide specificity in some cases seems to be similar to that identified for cAPK or MLCK, but in several cases the peptides that are phosphorylated are distinct from those recognised by other kinases. It has been found that an arginine three residues towards the N-terminal end of the polypeptide from the serine or threonine which accept the phosphate group seems to be essential. It has also been suggested that acidic residues within two residues of the phosphorylated amino acid or a hydrophobic residue on the C-terminal side of the phosphorylation site might also be important.

The other multifunctional Ca^{2+}/calmodulin-dependent protein kinases, include kinase I and kinase III. Kinase I, is a monomer of molecular mass of approximately 40 kDa. Kinase I and kinase III, an elongation factor-2 kinase, seem to have a much narrower substrate specificity than kinase II. Kinase I has a substrate specificity more closely related to cAPK than to kinase II.

A fourth Ca^{2+}/calmodulin-dependent kinase, kinase IV, is expressed mostly in brain tissue and the thymus, but has also been found in the spleen and testis. However it appears to be absent from several tissues studied. It has been purified from rat brain as two bands on sodium dodecyl sulphate–polyacrylamide gel electrophoresis (SDS-PAGE), of 65 and 67 kDa. Again, its activation involves autophosphorylation, although the exact physiological role for this kinase is not known.

G Protein-coupled Receptor Kinases

A class of kinases has been discovered that are involved in the down regulation of the receptors acting via the trimeric class of G proteins. These cytosolic kinases are known as GRKs, of which the best studied are probably βARK and rhodopsin kinase.

At least six GRKs are known, of which two are classed as βARKs. Towards the C-terminal ends of these proteins are PH domains capable of binding the βγ subunit complex from the trimeric G protein. Such an interaction between the G protein subunits and the kinase results in the translocation of the kinase to the membrane where it can interact with the receptor. With GRK2 and GRK3, lipids are also seen to stimulate this activity, including phosphatidylserine. However, the activity is inhibited by phosphatidylinositol 4,5-bisphosphate (PtdInsP$_2$) at high concentrations. It has been suggested that the lipids and the βγ G protein subunits compete for the same binding site on GRK-type kinases and that activation is actually dependent on the presence of a charged lipid.

Further to regulation by lipids and G proteins, βARK has been shown to be phosphorylated by PKC. Again, the translocation of the GRK to the membrane is enhanced along with an increase in kinase activity. Phosphorylation of the kinase is towards the C-terminal end of the polypeptide.

Of the GRKs not classed as βARKs, GRK4 is expressed most highly in the testis but is actually a family of four GRKs that arise from alternative splicing of the mRNA, involving exons II and XV. Meanwhile, GRK5 appears to be a unique member of the family of GRKs. It is capable of phosphorylation of rhodopsin, the M$_2$ muscarinic cholinergic receptor and the β$_2$-adrenergic receptor and is inhibited by heparin.

What is being recognised and phosphorylated by these kinases? In the β$_2$-adrenergic receptor, the phosphorylated residues are all located within a 40 amino acid stretch at the extreme C-terminal end of the receptor polypeptide. GRK5 phosphorylates threonines at positions 384 and 393 and serines at 396, 401, 407 and 411, while GRK2 phosphorylates serines 396, 401, 411 and threonine 384. Such a phosphorylation pattern is similar to that seen with rhodopsin. It should be noted that it is only the activated receptors which are phosphorylated, so the conformation of the protein must be important in kinase–receptor recognition.

The control of the receptor is thought to work in the following way. Once the receptor has bound ligand the G protein is activated. G$_\alpha$ transmits its signal to, for example, adenylyl cyclase, while the βγ complex activates the GRK. GRK phosphorylates only those receptors bound to ligand and deactivates the receptor, stopping further activation of G proteins. Further, phosphorylation of the receptor allows its interaction with a protein known as β-arrestin, which further inactivates the receptor. Therefore the receptor is desensitised and the cell shows adaptation to a prolonged exposure to ligand.

Haem-regulated Protein Kinase

It is not only enzymes in a metabolic pathway that are under the control of specific kinases; protein synthesis can also be regulated in this way. The classical example of this is the work done in reticulocytes, where it has been shown that globin synthesis is regulated by haem. It would be a waste for a cell to produce a large quantity of globin polypeptides if there is no protohaem available for completion of the holoenzyme. This mechanism is regulated by phosphorylation of one of the protein synthesis initiation factors, eIF-2, used to bring Met-tRNA to the ribosome to start the new polypeptide chain. If phosphorylated, this initiation factor forms a stable complex with guanine nucleotide exchange factor (GEF), preventing the exchange of guanosine diphosphate (GDP) for GTP on the initiation factor and so preventing another round of protein synthesis initiation.

This phosphorylation of eIF-2 is catalysed by a kinase known as haem-regulated protein kinase or haem-controlled repressor (HCR). The purified kinase has a molecular mass of approximately 95 kDa under denaturing conditions, but shows an apparent molecular mass of about 150 kDa under non-denaturing conditions. The sequence recognised by the kinase appears to be Leu-Leu-**Ser**-Glu-Leu-Ser, where the first serine is the site of phosphorylation. The only known substrate for the enzyme is initiation factor eIF-2 except that the enzyme also undergoes autophosphorylation. Both phosphorylation events are inhibited by the presence of haem. The activity of the kinase is also affected by the state of its sulphydryl groups: if these groups are oxidised or modified the enzyme is found in the active state even in the presence of haem. Other factors might also be involved in the activation of this kinase.

A similar control of translation is seen with an IFN-induced kinase, suggesting that phosphorylation might be a commonly used pathway for the control of protein synthesis.

Plant-specific Serine/Threonine Kinases

Although the text above has been devoted to the discussion of kinases as characterised in animal tissues, it should be noted that plants and fungi also use protein phosphorylation in the control of cellular functions, and in fact plants have been found to contain several kinases very similar to those described above. These are controlled by mechanisms like those described above. However, several plant-specific kinases have also been identified and partially or completely purified. One of the best characterised is a kinase that phosphorylates the light-harvesting chlorophyll *a/b* complex (LHC). LHC is phosphorylated on at least one threonine residue, activating it when exposed to low levels of red light. The kinase, found in thylakoid membranes, has an approximate molecular mass of 64 kDa and its activation probably involves the redox state of plastoquinone.

With growing evidence that plants also contain kinases analogous to those found in animal tissues, one of the exciting areas of research is the search for animal-like proteins and control mechanisms in plants. For example, several effects of phorbol esters have been seen in plant tissues, suggesting the presence of a PKC-like enzyme, which is yet to be identified.

Tyrosine Kinases

The second major class of protein kinases includes those that add a phosphoryl group to tyrosine as opposed to serine or threonine. Phosphoryl addition to tyrosine is much less common than modification on serine or threonine: an analysis in 1980 of the phosphoamino acid content of a cell revealed that only about 0.05% was phosphotyrosine. However, this should not belittle their importance. The tyrosine kinases are themselves found in two broad groups, those which are soluble and those which are part of a receptor.

The general phosphorylation site has been characterised and includes a lysine or arginine residue seven amino acids to the N-terminal side of the tyrosine. An acidic amino acid, such as aspartate or glutamate, is quite often found three or four residues to the N-terminal side of the tyrosine, giving the consensus sequence:

-[Lys/Arg]-X-X-[Asp/Glu]-X-X-X-Tyr-
-[Lys/Arg]-X-X-X-[Asp/Glu]-X-X-Tyr-   or

As with most of these signatures, there are exceptions that do not seem to fit into this neat pattern.

Receptor Tyrosine Kinases

The receptors comprising this broad class of kinases can be split into 14 groups, characterised by their general structural patterns although they all share the same basic topology. They all possess an extracellular ligand-binding domain, a single transmembrane domain and a cytoplasmic domain that contains the kinase activity. These receptors are also discussed in Chapter 3.

In all cases, binding of the ligand to the extracellular binding site of the receptor activates the kinase activity on the cytoplasmic side of the receptor. Phosphorylation by this kinase activity then leads to intracellular signalling by the activation of common signalling pathways. Enzymes activated by this type of kinase include, amongst many others, phosphatidylinositol 3-kinase involved in the inositol pathway, GTPase-activating protein, involved in G-protein signalling, and MAP kinases.

Not only does ligand binding turn on tyrosine kinase activity towards the proteins in the signalling pathway, but these receptor kinases also autophosphorylate and in many cases the receptors also dimerise, for example the EGF receptor family of kinases. Where the receptor exists in different isoforms, the dimers can be made up of two of the same polypeptide or may be a mixture of isoforms. An example of this is seen with the PDGF receptor, which can have dimers comprising αα, ββ or αβ, probably depending on the precise ligand that has been bound.

In autophosphorylation, it is tyrosine residues on the cytoplasmic side of the receptor that are the acceptors for phosphoryl groups. This autophosphorylation may be intramolecular, where the polypeptide chain phosphorylates itself or, in some cases, intermolecular, where one polypeptide in a dimer phosphorylates the other and vice versa. This latter case is seen with the FGF receptor family. It is thought that autophosphorylation is the signal for binding of other cytoplasmic proteins, the phosphotyrosine residues creating a selective binding site. These observations have arisen from the use of synthetic peptides, which can be used to block the interactions of polypeptides by binding to recognition sites and stopping the binding of the true ligand. These studies have shown that amino acids other than phosphotyrosine are also involved in polypeptide interactions. For example, for the binding of phosphatidylinositol 3-kinase a methionine residue is required three residues to the C-terminal side of the phosphotyrosine. Residues to the N-terminal side of the tyrosine seem to be less important than those on the C-terminal side.

The importance of the creation of these binding sites by autophosphorylation may well be to generate high-affinity binding sites for the proteins to be phosphorylated, allowing efficient phosphorylation of proteins that might be in very low abundance within the cell. Once bound to the receptor protein kinase the bound protein will itself be phosphorylated on tyrosine, so altering its activity.

These polypeptides that bind RTKs not only rely on the presence of the phosphotyrosine residues of the receptor but themselves contain binding domains. The most common ones found are the SH2 domains. These domains are homologous to the non-catalytic region of the Src protooncoprotein, as discussed in Chapter 1. In some instances binding of the protein to the receptor is facilitated by the presence of a linker or adaptor protein, which quite commonly contains SH2 or SH3 domains. Examples of such adaptor proteins are GRB2 in mammals or Drk in the fly *Drosophila*. These proteins are often associated with GNRPs and the interaction of the adaptor with the activated receptor through SH2 domains and newly formed phosphotyrosine residues leads to the association of GNRP with membrane proteins and ultimately activation of its respective G protein. Such a scheme is illustrated in Figure 4.8 and is discussed further in Chapter 9.

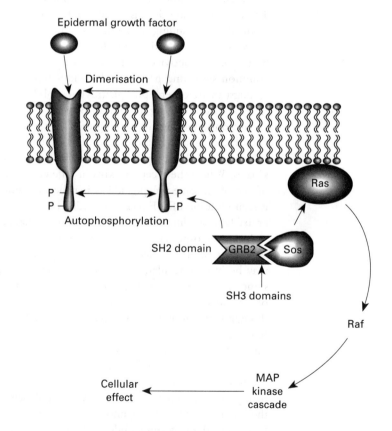

Figure 4.8
Signalling cascade associated with epidermal growth factor detection as an example of signalling from receptor tyrosine kinases. Binding of the growth factor to the receptor causes dimerisation and autophosphorylation of the receptor. The phosphotyrosines formed create binding sites for the SH2 domains of the GRB2 adaptor protein. This protein is found associated with Sos, a guanine nucleotide releasing protein, through SH3 domains. The new receptor–GRB2–Sos complex causes activation of the G protein Ras, which leads to the activation of a kinase cascade and ultimately the cellular effect. MAP, mitogen-activated protein.

In several classes of RTKs common areas of homology have been characterised, which will be useful in the identification of new kinases and assigning kinase activity to proteins that may at present have no function assigned to them.

In many of these receptor kinases, it is not only tyrosine that can be phosphorylated but also serine or threonine. In the insulin receptor family, serine/threonine phosphorylation occurs on the β subunit of the receptor as well as on a number of intracellular substrates such as the protein kinase Raf-1. However, the exact significance of this is yet to be determined.

Cytosolic Tyrosine Kinases

Kinases that contain tyrosine phosphorylation capacity are not only membrane bound, as the receptor kinases, but may also be soluble and reside in the cytoplasm of the cell. One such group of kinases are the JAKs, named after Janus in Roman mythology, who was the god of doors and gateways and signified beginnings that ensured good endings. The statue of Janus has two heads, which are seen to be gazing in opposite directions. JAKs contain tandem but non-identical catalytic domains, but they lack both SH2 and SH3 domains. They are involved in the signal-transduction pathways from cytokines, particularly in leucocytes and lymphocytes. They associate with an activated cytokine receptor, which has undergone a conformational change on binding to its ligand; hence, once activated themselves, JAKs cause the phosphorylation of their respective receptor protein as well as cellular proteins such as STAT proteins, which leads to the activation of transcription (Figure 4.9). As these cytosolic STAT proteins contain SH2 domains, once they are phosphorylated by JAKs they can dimerise by means of the association of the new phosphotyrosine of one STAT protein with the SH2 domain of another. Such activated dimers then cause the alteration of transcription observed.

Members of the JAK family include JAK1, JAK2 and Tyk2, and now JAK3. Different members of the family have been implicated in message transfer from different receptors. For example, activation of the erythropoietin receptor leads to activation of JAK1, while JAK1 associates with JAK2 in the IFN-γ transduction pathway. JAK3 has been implicated in the signalling from IL-2 and IL-4. This latter kinase is a polypeptide of approximately 120 kDa, although it probably exists as at least three variants. JAK3 is slightly different from other members of the JAK family in being slightly smaller and having a tissue-specific expression pattern.

Mitogen-activated Protein Kinases

Many external factors, such as cytokines and growth factors, can lead to the phosphorylation and activation of a family of kinases that have only recently been discovered. These are the extracellular signal-regulated kinases (ERKs), more commonly known as MAP kinases (MAPK). Activation of these serine/threonine kinases can lead to their translocation to the nucleus and phosphorylation and activation of transcription factors, causing promoted growth and differentiation for example.

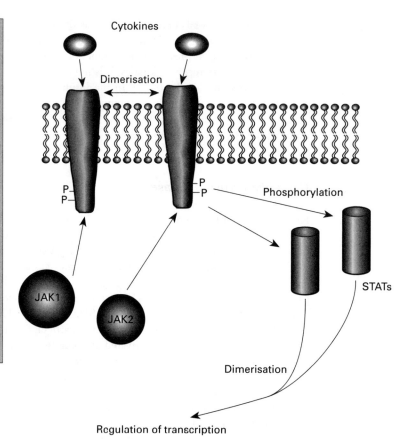

Figure 4.9
Signalling cascade proposed to lead from cytokine receptors. Binding of the cytokine, such as interferon–γ leads to dimerisation of the receptor and the recruitment of soluble tyrosine kinases, Janus kinases (JAKs). Phosphorylation of the receptors, JAKs and other proteins (STATs signal transducers and activators of transcription) takes place. STAT proteins themselves dimerise and carry the signal to the nucleus, where transcription is regulated.

These kinases are themselves activated by phosphorylation in an unusual manner: phosphorylation on tyrosine and threonine through the action of a specific threonine/tyrosine kinase known as MAPK kinase (MAPKK). This enzyme is alternatively known as MAPK/ERK kinase (MEK). The sequence commonly recognised and phosphorylated by MAPKK is Thr-Glu-Tyr, where both the threonine and tyrosine residues receive phosphoryl groups leading to the activation of MAPK. However, the glutamate residue may be variant in some cases.

MAPKK is itself also regulated by phosphorylation, catalysed this time by MAPKK kinase, otherwise known as MAPKKK or MEK kinase (MEKK). This is in fact a group of related serine/threonine kinases. One of the proteins that catalyses this phosphorylation is Raf. The MAPKKKs (MEK kinases) from mammalian systems show conservation of the catalytic domains but interestingly they show divergence of a regulatory domain towards the N-terminus. A typical MAPK cascade is illustrated in a simplified form in Figure 4.10. The initial signal for this cascade of phosphorylations can originate in the activation of a RTK activity or, in lower organisms, in the activation of a two-component system as described below. Activation of the MAPK cascade can also involve the activation of G proteins, in some cases through the action of G_α, in other cases through the action of $G_{\beta\gamma}$ and

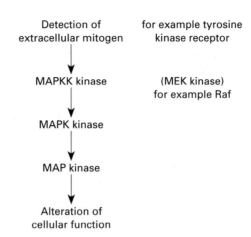

Figure 4.10
Simplified schematic representation of the mitogen-activated protein (MAP) kinase cascade.

also through Ras, while MAPKs have been shown to be activated by reactive oxygen species, as discussed in Chapter 8.

The activation by EGF serves as a good example (see Figure 4.8). The RTK leads to formation of binding sites for, and the activation of, an adaptor protein, GRB2, which contains both SH2 and SH3 domains. The SH2 domains are responsible for the adaptor protein binding to the receptor once it has been autophosphorylated on tyrosine residues, while the SH3 domains are involved in its interaction with a GNRP and the subsequent activation of Ras. Once Ras has been converted to the GTP-bound form it is able to cause the activation of the kinase Raf and phosphorylation catalysed by Raf leads to activation of the rest of the cascade.

There is also the possibility that the kinase Raf can be activated by PKC, which may itself be activated by the action of PLC, known to be acted upon by trimeric G proteins. Another example of a transduction cascade involving MAPKs is the insulin cascade, as discussed in Chapter 9.

It has been postulated that other proteins are also involved in the MAPK pathway, some of which are themselves kinases but others that are proposed to have a function as scaffold proteins rather than possessing kinase activity. Such a protein is STE5 from *Saccharomyces cerevisiae*, which contains a zinc finger-like conformational domain and has been shown to interact with other components of the MAPK pathway. Although the same components have been seen to function in more than one pathway and more than one pathway seems to have the same end-result in some cases, the existence of such a scaffold structure would be an attractive hypothesis to explain the relative lack of interaction between different MAPK pathways in the same cell. At least six distinct MAPK pathways have been identified in yeast and it is difficult to see how the signals remain distinct with such similar and related components. The existence of a physical scaffold and kinase complexes would help explain these observations.

Once the MAPK has been phosphorylated and therefore activated it must have a role in phosphorylating further proteins, causing an alteration in their activity. Transcription factors are known to be targets for MAPK, with phosphorylation causing an increase in the rate of transcription of specific genes, for example in the pheromone response of yeast.

Like all signalling pathways, once turned on the response has to be turned off again. In this case, the reversal is effected by dephosphorylation catalysed by phosphatases. A group of phosphatases whose synthesis is inducible by growth factors and stimuli of the MAPK pathway has been shown to be involved in switching off the pathway. These phosphatases seem to be specific for MAPK, while amino acid sequences appear to have some homology to a dual-specificity tyrosine/serine phosphatase from vaccinia virus, VH1. One such phosphatase, CL100, also shows some homology to another dual-specificity phosphatase, cdc25. Another phosphatase, PAC-1, has been found to be localised to the nucleus, where MAPK has its effect.

Other likely candidates involved in the cessation of the MAPK cascade are less specific phosphatases, such as PP2A or CD45.

Histidine Phosphorylation

Although the most common phosphorylations of proteins are based on the amino acids serine, threonine or tyrosine, they are not the only amino acids to participate in this type of reaction in order to cause an alteration in the activity of the protein and hence the propagation of a cellular signal. Histidine and aspartate are found to be phosphorylated in bacteria. These are relatively high-energy phosphoamino acids and are involved in what has been termed the two-component signalling system. Detection of a stimulus leads to phosphorylation of the detector protein on a histidine residue, using ATP, and the phosphoryl group is subsequently transferred to a second protein, the response regulator, which becomes phosphorylated on an aspartate residue. These events lead to an alteration in the function of this latter protein, for example it may have enhanced DNA-binding capacity and so alter DNA transcription rates.

In general, the first phosphorylation step is an autophosphorylation event. Histidine kinases belong to a group of kinases that have several conserved domains in common. Five main domains have been identified but a kinase may not contain all of them and may contain other functional domains. Most contain a region in which there is a conserved histidine that receives the phosphoryl group, but sometimes the autophosphorylation may take place in a separate region of the polypeptide. Two glycine-rich regions, termed G1 and G2, contain kinase activity as well as phosphatase activity and nucleotide-binding ability; two other regions, termed N and F, are conserved but have no assigned function. The stimulus can be detected either by a receptor region that is part of the kinase polypeptide or by a separate polypeptide, which activates the kinase by protein–protein interaction. An integral receptor domain would be extracellular, with the polypeptide having a transmembrane region and the other domains on the intracellular side of the membrane. Other histidine kinases are soluble and are found in the cytoplasm of the cells. These histidine kinases are classical examples of modular proteins, as different domains seem to be glued together to give the catalytic activity and specificity required.

The response regulators, which receive the phosphoryl group on their aspartate residues, share a common domain containing two aspartate residues and a conserved lysine residue. Phosphorylation leads to an alteration of the protein's function. The protein can return to the inactive form by an intrinsic phosphatase activity or by the activity of a separate phosphatase enzyme.

Systems using this two-component type of signalling pathway include those which monitor osmolarity, temperature, pH and cell density. In fact, over 50 such systems have been identified.

Some bacterial proteins now coming to light contain both the histidine kinase domains and the conserved regions usually associated with the response regulator proteins. These proteins are referred to as hybrid kinases.

Using the conserved sequences of histidine kinases as a basis, researchers have looked for such kinases in higher organisms. Such a strategy has revealed sequences homologous to histidine kinases in the fungi *Neurospora crassa*, the slime mould *D. discoideum* and the yeast *S. cerevisiae*, while a protein sharing homology to a hybrid histidine kinase has been found in the higher plant *Arabidopsis thaliana*. It may be that histidine and aspartate phosphorylations will be discovered to be widespread and abundant in eukaryotic systems.

Phosphatases

No treatment of phosphorylation would be complete without a discussion on how the phosphoryl group is removed to reverse the cycle and return the protein to its previous state, be that active or inactive. This process of dephosphorylation is carried out by a group of enzymes called phosphatases. These again are split broadly into two groups: those which remove the phosphoryl group from serine or threonine residues, the serine/threonine phosphatases, and ones which remove the phosphate groups from tyrosine residues, the tyrosine phosphatases. In a similar fashion to the kinases, the number of phosphatases coded for in the human genome has been estimated, research suggesting that it could be as many as 1000. As with the kinases, plants and fungi contain phosphatases analogous to those characterised in animals but some plant-specific phosphatases have also been found, such as the LHC phosphatase associated with the thylakoid membranes.

Serine/Threonine Phosphatases

Several different types of these enzymes have been identified, referred to as PP1, PP2A, PP2B, PP2C, PP4 and PP5. Interestingly, it has been estimated that the total cellular serine/threonine phosphatase activity is approximately equal to the total cellular serine/threonine kinase activity, suggesting a balance. Indeed, the cellular concentrations of the major forms of these two enzymes are also comparable.

PP1 has a broad substrate specificity and in mammals two closely related isoforms have been identified, PP1α and PP1β. These two enzymes arise from alternative splicing of the same gene. Two thermostable proteins are able to inhibit the activity of PP1: inhibitor 1 and inhibitor 2. PP1 appears to be involved in the regulation of glycogen metabolism, where it is associated with a 160-kDa glycogen-binding protein. This glycogen-binding protein is itself under the control of phosphorylation. Phosphorylation of this latter protein by cAPK makes it unable to bind to the catalytic subunit of the phosphatase, showing that PP1 is under the control of the cAMP-signalling pathway. Inhibitor 1 is also only active in inhibiting PP1 when it is in the phosphorylated state, again effected by cAPK. Therefore the action of the cAMP pathway is twofold: deactivation of PP1 and thus prevention of dephosphorylation of proteins phosphorylated by kinases that are turned on directly or indirectly by the rise in cAMP.

PP2A is a trimeric protein and has some similarity to PP1. PP2A has two identified isoforms of the catalytic subunit, PP2Aα and PP2Aβ, but in this case they are separate gene products. The other subunits involved are a regulatory subunit of 65 kDa and a third subunit which shows a degree of variability. PP2B, also known as calcineurin, is a Ca^{2+}-dependent enzyme. In the presence of Ca^{2+}/calmodulin the activity of PP2B is increased. It exists as a heterodimer of a catalytic subunit (A subunit) and a Ca^{2+}-binding subunit (B subunit), which contains four Ca^{2+} binding regions. Therefore, the effect of Ca^{2+} may be direct or through the action of calmodulin. One form of the enzyme that has been purified has a molecular mass of approximately 80 kDa, composed of a 60-kDa A subunit and a 20-kDa B subunit. However, the A subunit seems to be quite variable between tissues and may account for reported differences in substrate specificity. The subcellular location of the enzyme may be dictated by the B subunit, which may be myristoylated, allowing its association with membranes.

PP2C, a 42–45-kDa monomeric protein, appears to be a Mg^{2+}-dependent enzyme, while PP5 is primarily found in the nucleus and contains sequences reminiscent to those found with proteins associated with RNA and DNA binding.

Studies of the roles of phosphatases, like that of many other signalling components, has been greatly enhanced by the use of inhibitors. PP1 and PP2A have been shown to be strongly inhibited by okadaic acid, a fatty acid produced by dinoflagellates. It has very little effect on PP2B, and has no effect on PP2C or tyrosine phosphatases. A similar inhibitory pattern is seen with tautomycin from *Streptomyces*.

Tyrosine Phosphatases

In 1988 when Tonks and colleagues published the partial sequence of the first protein tyrosine phosphatase (PTP 1B), it was found that not only did the enzyme lack homology with serine/threonine phosphatases but another such enzyme had been cloned a few years earlier, although that protein, CD45, had been assigned no function. It is now known that there are two main classes of PTPs: those which are intracellular and those which are receptor linked, as is seen with tyrosine kinase families.

PTP 1B can be thought of as the model for the family of intracellular tyrosine phosphatases. The N-terminal end of the single polypeptide contains the catalytic domain, while the C-terminal end contains a signal to direct the intracellular location of the enzyme. The enzyme appears to be localised to the cytoplasmic side of the endoplasmic reticulum, but removal of the C-terminal end relocates the enzyme into the cytosol. Relocation of the enzyme also alters its functional activity and it is thought that regulation of the phosphatase may be through regulation of its location in the cell. Alternative splicing of the mRNA for some PTPs might well be responsible for controlling their cellular location.

On to this basic structure for intracellular PTPs are added additional features (Figure 4.11), for example PTP 1C also contains two protein-binding SH2 domains.

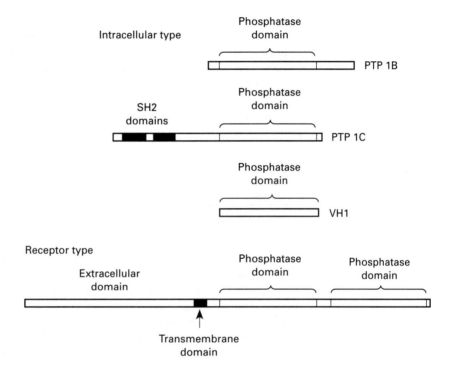

Figure 4.11
Domain structure of several protein tyrosine phosphatases.

As has been mentioned above for the RTKs, these domains might be used to locate and bind phosphotyrosine residues, so aiding the catalytic action of the enzyme. The SH2 domains might also be crucial to maintain the enzyme's intracellular location.

Another identified addition to the basic PTP 1B structure is the addition of a domain that resembles cytoskeletal-associated proteins. For example, PTPMEG1 has an area which shares some homology to erythrocyte protein band 4.1. Again, this is probably important in the maintenance of intracellular location of the phosphatase. Tyrosine phosphorylation is thought to be important in the polymerisation of tubulin, a main element of the cytoskeleton, and PTPs would be vital in this regulatory mechanism.

The smallest PTP identified contains the phosphatase domain and nothing else. This protein, VH1, is coded for by the vaccinia virus. Interestingly this phosphatase also shows activity to serine-phosphorylated sites as well as tyrosine-phosphorylated sites.

Just as PTP 1B can be seen as a model for intracellular PTPs, CD45 can be used as a model for receptor-linked PTPs. CD45 is a glycoprotein, with O-linked glycosylation in the N-terminal extracellular domain. Unlike the intracellular PTPs, the receptor-linked family nearly all have two phosphatase domains, although exceptions to this have been found (see Figure 4.11). However, using mutations of the amino acids thought to be crucial in the catalytic action of this phosphatase (mainly a conserved cysteine residue), it is thought that the second catalytic site is probably inactive in removing the phosphoryl group from tyrosines. However, it could have a function in bringing about the association of the enzyme with its substrate, in a similar manner to the SH2 domains seen elsewhere.

PTPs, like the tyrosine kinases, are probably regulated by phosphorylation, although the precise mechanisms have yet to be identified.

Several inhibitors have been identified that are useful in identifying the activity of PTPs. Molybdate and particularly orthovanadate are inhibitors of all PTPs, while other compounds have been used as diagnostic tools for certain PTPs. Such molecules include zinc ions, ethylenediaminetetraacetic acid (EDTA) and spermine.

Other Covalent Modifications

Although the discussion above has concentrated on the covalent addition of the phosphoryl group and its subsequent removal from a polypeptide as means of regulating activity, it is not the only covalent modification seen in the control of enzyme function. Many proteins are proteolytically cleaved, turning them from an inactive precursor into an active enzyme. The classical case here is seen in the blood-clotting cascade, where one factor cleaves and activates another, which then goes on to cleave the next and so on, leading to great amplification of the initial signal. However, unlike phosphorylation, such reactions are not reversible as once the polypeptide has been cleaved it is not rejoined. An example of relevance to cell signalling is the cleavage activation of PKC.

A more reversible covalent modification is adenylation. The adenylyl group (AMP) from ATP is added to a target tyrosine residue of the polypeptide, releasing inorganic pyrophosphate, the breakdown of which probably helps to drive the reaction. Reversal releases the AMP, restoring the enzyme to its former condition. An example of this is seen in *E. coli* where glutamine synthase in the cell's nitrogen metabolism is controlled in this way. Adenylation inhibits the enzyme's activity.

Many proteins are covalently modified by the addition of a myristoyl group. This is a 14-carbon fatty acid that is added to the N-terminal end of the polypeptide. The donor for myristoylation is myristoyl CoA, while the enzyme that catalyses the reaction is *N*-myristoyltransferase. For this reaction to take place, the target polypeptide must have a glycine residue at its N-terminal end with a typical consensus sequence being:

-Gly-X-X-X-[Ser/Thr]-Y-Y-

where X is a variety of amino acids but Y is a basic amino acid.

Palmitoylation involves the addition of a palmityol group, a 16-carbon fatty acid, donated from palmitoyl CoA. This is added to a cysteine residue. An example of a protein that has undergone this type of modification is rhodopsin.

Other covalent modifications include farnesylation, which is the addition of a farnesyl group, a 15-carbon fatty acid. The donor in this case is farnesyl pyrophosphate. This process takes place at the C-terminal end of the polypeptide where a consensus sequence of Cys-Y-Y-X is found. Cys is the target for the modification, Y would be aliphatic in nature, while X could be anything. Typically, the Y-Y-X sequence would be subsequently removed and the end of the polypeptide would also be methylated. An example of a signalling protein altered in this way is the G protein Ras.

The last modification considered here is the addition of the 20-carbon fatty acid geranylgeranyl, added in a process called geranylgeranylation. Again, like farnesylation, the pyrophosphate form is the source of the fatty acid and again it is attached to the C-terminal end of the polypeptide. However, the consensus sequences needed at the C-terminus are Cys-Cys, Cys-Cys-X-X or Cys-X-Cys. One or both cysteine groups could be modified. An example of a protein modified in this way is the monomeric G protein Rab.

These additions of fatty acid groups to protein are important in the association of these proteins with membranes. For example, if Ras is not able to be modified by attachment of a farnesyl group it is not able to associate properly with the plasma membrane and is unable to function in signal cascades.

Plants also often use the reversible reduction of cystine disulphide bridges commonly in their regulation of enzyme activity. Thioredoxin is used as the reductant in the reaction, the re-reduction of thioredoxin being light driven.

Summary

Most signalling pathways, at one or more points, undergo alteration of the activity of a protein's function, brought about by the addition of one or more phosphoryl groups to certain amino acid groups on the polypeptide chain. The importance of this covalent modification is highlighted in Figure 4.12. Such phosphorylation events are either classified as serine/threonine phosphorylations or tyrosine phosphorylations, depending on the target amino acids. The enzymes that add phosphoryl groups are known as kinases, with the reverse reaction being catalysed by phosphatases. The human genome probably contains over 1000 genes for each type of enzyme.

Serine/threonine kinases include PKC, cAPK, Ca^{2+}/calmodulin-dependent protein kinase and cGPK, which in general have wide substrate specificities, as well as more specific kinases such as phosphorylase kinase. cAPK exists in an inactive tetrameric state which dissociates on activation, i.e. on binding to cAMP, to release two catalytic subunits. PKC, which is activated by the presence of Ca^{2+} ions, is in fact two families of protein kinases, each family containing several members, which probably have subtle differences in both their activation and specificity. Another

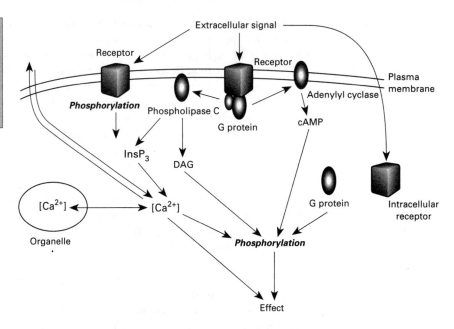

Figure 4.12
Schematic representation of the central role of phosphorylation in cell signalling. DAG, diacylglycerol; InsP$_3$, inositol trisphophate.

group of Ca^{2+}-controlled kinases are the Ca^{2+}/calmodulin-dependent kinases, again with several isoforms.

The other major class of kinases are the tyrosine kinases. These include the receptor kinases, which are membrane-bound receptors. Binding to their ligand results commonly in autophosphorylation of the receptor along with phosphorylation of intracellular proteins. Other tyrosine kinases include the soluble JAK family.

A recently discovered group of sequential kinases are those of the MAPK cascades. These pathways lead commonly from receptors that bind growth factors or insulin, and may involve the G protein Ras and the kinase Raf at the early part of the transduction pathway.

The two-component system of prokaryotes unusually involves the attachment of phosphoryl groups to histidine and aspartate. Such systems may be more widespread than first thought.

The reversal of kinase action is effected by phosphatases and again these are usually either serine/threonine specific or tyrosine specific, the latter being either cytosolic or receptor linked. Several isoforms of each are known to exist.

Further reading

General

Edelman, A. M., Blumenthal, D. A. and Krebs E. G. (1987) Protein serine/threonine kinases. *Annual Review of Biochemistry*, **56**, 567–613.

Hanks, S. K. and Quinn A. M. (1991) Protein kinase catalytic domain sequence database: identification of conserved features of primary structure and classification of family members. *Methods in Enzymology*, **200**, 38–62.

Hanks, S. K., Quinn, A. M. and Hunter, T. (1988) The protein kinase family: conserved features and deduced phylogeny of the catalytic domains. *Science*, **241**, 42–52.

Hidaka, H. and Kobayashi, R. (1994) Protein kinase inhibitors. *Essays in Biochemistry*, **28**, 73–97.

Hunter, T. (1991) Protein kinase classification. *Methods in Enzymology*, **200**, 3–37.

Hunter, T. (1995) Protein kinases and phosphatases: the yin and yang of protein phosphorylation and signalling. *Cell*, **80**, 225–36.

Kamm, K. E. and Stull, J. T. (1989) Regulation of smooth muscle contractile elements by second messengers. *Annual Review of Physiology*, **51**, 299–313.

Kikkawa, U., Kishimoto, A. and Nishizuka, Y. (1989) The protein kinase C family: heterogeneity and its implications. *Annual Review of Biochemistry*, **58**, 31–44.

Kishimoto, A., Nishiyama, K., Nakanishi, H. *et al.* (1985) Studies on the phosphorylation of myelin basic protein by protein kinase C and adenosine 3′,5′-monophosphate-dependent protein kinase. *Journal of Biological Chemistry*, **260**, 12492–9.

Knighton, D. R., Zheng J., Ten Eyck L. F. *et al.* (1991) Crystal structure of the catalytic subunit of cyclic adenosine monophosphate-dependent protein kinase. *Science*, **253**, 407–14.

Ryazanov, A. G., Rudkin, B. B. and Spirin, A. S. (1991) Regulation of protein synthesis at the elongation stage: new insights into the control of gene expression in eukaryotes. *FEBS Letters*, **285**, 170–5.

Stull, J. T. (1988) Myosin light chain kinases and caldesmon: biochemical properties and role in skeletal smooth muscle contractions. *Calmodulin*, Cohen, P. and Klee, C. (eds), pp. 91–122. Elsevier, Amsterdam.

Sun, H. and Tonks, N. K. (1994) The co-ordinated action of protein tyrosine phosphatases and kinases in cell signalling. *Trends in Biochemal Sciences*, **19**, 480–5.

Taylor, S. S., Buechler, J. A. and Yonemoto, W. (1990) cAMP-dependent protein kinase: framework for a diverse family of regulatory enzymes. *Annual Review of Biochemistry*, **59**, 971–1005.

Woodget J. R., Gould, K. L. and Hunter, T. (1986) Substrate specificity of protein kinase C: use of synthetic peptides corresponding to physiological sites as probes for substrate recognition requirements. *European Journal of Biochemistry*, **161**, 177–84.

cAMP-dependent Protein Kinase

Bubis, J., Vedvick, T. S. and Taylor, S. S. (1987) Antiparallel alignment of the two protomers of the regulatory subunit dimer of cAMP-dependent protein kinase I. *Journal of Biological Chemistry*, **262**, 14961–6.

Carr, S. A., Biemann, K., Shoji, S., Parmlee, D. C. and Titani, K. (1982) Normal tetradecanyol is the NH_2-terminal blocking group of the catalytic subunit of cyclic AMP dependent protein kinase from bovine cardiac muscle. *Proceedings of the National Academy of Sciences USA*, **79**, 6128–32.

Døskeland, S. O. and Øgreid, D. (1981) Binding proteins for cyclic AMP in mammalian tissues. *International Journal of Biochemistry*, **13**, 1–19.

Flockhart, D. A. and Corbin, J. D. (1982) Regulatory mechanisms in the control of protein kinases. *CRC Critical Reviews in Biochemistry*, **12**, 133–86.

Glass D. B., El-Maghrabi, M. R. and Pilkis, S. J. (1986) Synthetic peptides corresponding to the site phosphorylated in 6-phosphofructo-2-kinase/fructose-2,6, -bisphosphatase as substrates of cyclic nucleotide-dependent protein kinase. *Journal of Biological Chemistry*, **261**, 2987–93.

Hanks, S. K., Quinn, A. M. and Hunter, T. (1988) The protein kinase family: conserved features and deduced phylogeny of the catalytic domains. *Science*, **241**, 42–52.

Øgreid, D. and Døskeland, S. O. (1981) The kinetics of the interaction between cyclic AMP and the regulatory moiety of protein kinase II. *FEBS Letters*, **129**, 282–6.

Reed, J., Kinzel, V., Kemp, B. E., Cheung, H.-C. and Walsh, D. A. (1985) Circular dichroic evidence for an ordered sequence of ligand/binding site interaction in the catalytic reaction of the cAMP-dependent protein kinase. *Biochemistry*, **24**, 2967–73.

Slice, L. W. and Taylor, S. S. (1989) Expression of the catalytic subunit of cAMP dependent protein kinase in *Escherichia coli*. *Journal of Biological Chemistry*, **264**, 20940–6.

Smith, S. B., White, H. D., Siegel, J. and Krebs, E. G. (1981) Cyclic AMP-dependent protein kinase I: cyclic nucleotide binding, structural changes and release of the catalytic subunits. *Proceedings of the National Academy of Sciences USA*, **78**, 1591–5.

Takio, K., Smith, S. B., Krebs, E. G., Walsh, K. A. and Titani, K. (1984) Amino acid sequence of the regulatory subunit of bovine type II adenosine cyclic 3'-5'-phosphate dependent protein kinase. *Biochemistry*, **23**, 4200–6.

Takio, K., Wade, R. D., Smith, S. B., Krebs, E. G., Walsh, K. A. and Titani, K. (1984) Guanosine cyclic 3',5' phosphate protein kinase: a dimeric protein homologous with two separate protein families. *Biochemistry*, **23**, 4207–18.

Taylor, S. S., Buechler, J. A., Slice, L. *et al.* (1989) cAMP dependent protein kinase: a framework for a diverse family of enzymes. *Cold Spring Harbor Symposia on Quantitative Biology*, **53**, 121– 30.

Taylor, S. S., Buechler, J. A. and Knighton, D. (1990) cAMP-dependent protein kinase: mechanism for ATP protein phosphotransfer. *CRC Critical Reviews in Biochemistry*, 1–41.

Titani, K., Sasagawa, T., Erisson, L. H. *et al.* (1984) Amino acid sequence of the regulatory subunit of bovine type I adenosine cyclic 3'-5'-phosphate dependent protein kinase. *Biochemistry*, **23**, 4193–9.

Zick, S. K. and Taylor, S. S. (1982) Interchain disulphide bonding in the regulatory subunit of cAMP dependent protein kinase I. *Journal of Biological Chemistry*, **257**, 2287–93.

cGMP-dependent Protein Kinase

Corbin, J. D., Øgreid, D., Miller, J. P., Suva, R. H., Jastorff, B. and Døskeland, S. O. (1986) Studies of cGMP analog specificity and function of the two intrasubunit binding sites of cGMP-dependent protein kinase. *Journal of Biological Chemistry*, **261**, 1208–14.

Lincoln, T. M. and Corbin, J. D. (1983) Characterisation and biological role of the cGMP dependent protein kinase. *Advances in Cyclic Nucleotide Research*, **15**, 139–92.

Takio, K., Smith, S. B., Walsh, K. A., Krebs, E. G. and Titani, K. (1983) Amino acid sequence around a hinge region and its autophosphorylation site in bovine lung cGMP dependent protein kinase. *Journal of Biological Chemistry*, **258**, 5531–6.

Takio, K., Wade, R. D., Smith, S. B., Krebs, E. G., Walsh, K. A. and Titani, K. (1984) Guanosine cyclic 3′-5′-phosphate dependent protein kinase: a dimeric protein homologous with two separate protein families. *Biochemistry*, **23**, 4207–18.

Protein Kinase C

Coussens, L., Parker, P. J., Rhee, L. *et al.* (1986) Multiple, distinct forms of bovine and human protein kinase C suggest diversity in cellular signalling pathways. *Science*, **233**, 859–66.

Coussens, L., Rhee, P. J., Parker, P. J. and Ullrich, A. (1987) Alternative splicing increases the diversity of the human protein kinase C family. *DNA*, **6**, 389–94.

House, C. and Kemp, B. E. (1987) Protein kinase C contains a pseudosubstrate prototype in its regulatory domain. *Science*, **238**, 1726–8.

Hubbard, S. R. (1995) X-ray absorption spectroscopy studies of protein kinase C. *Methods in Enzymology*, **252**, 123–32.

Kikkawa, U. and Nishizuka, Y. (1986) The role of protein kinase C in transmembrane signalling. *Annual Review of Cell Biology*, **2**, 149–78.

Kosaka, Y., Ogita, K., Ase, K., Nomura, H., Kikkawa, U. and Niskizuka, Y. (1988) The heterogeneity of protein kinase C in various rat tissues. *Biochemical and Biophysical Research Communications*, **151**, 973–81.

Kubo, K., Ohno, S. and Suzuki, K. (1987) Primary structures of human protein kinase C beta I and beta II differ only in their C-terminal sequences. *FEBS Letters*, **223**, 138–42.

Naor, Z., Shearman, M. S., Kishimoto, A. and Nishizuka, Y. (1988) Calcium independent activation of hypothalamic type I protein kinase C by unsaturated fatty acids. *Molecular Endocrinology*, **2**, 1043–8.

Nishizuka, Y. (1986) Studies and perspectives of protein kinase C. *Science*, **233**, 305–12.

Nishizuka, Y. (1988) The molecular heterogeneity of protein kinase C and its implication for cellular regulation. *Nature*, **334**, 661–5.

Ono, Y., Fujii, T., Ogita, K., Kikkawa, U., Igarashi, K. and Nishizuka, Y. (1987) Identification of three additional members of rat protein kinase C family: δ, ε and ζ subspecies. *FEBS Letters*, **226**, 125–8.

Ono, Y., Kikkawa, U., Ogita, K. *et al.* (1987) Expression and properties of two types of protein kinase C: alternative splicing from a single gene. *Science*, **236**, 1116–20.

Ono, Y., Fujii, T., Ogita, K., Kikkawa, U., Igarashi, K. and Niskizuka, Y. (1988) The structure, expression and properties of additional members of the protein kinase C family. *Journal of Biological Chemistry*, **263**, 6927–32.

Parker, P. J., Coussens, L., Totty, N. *et al.* (1986) The complete primary structure of protein kinase C, the major phorbol ester receptor. *Science*, **233**, 853–9.

Ca²⁺/Calmodulin-dependent Protein Kinases

Ahmad, Z., DePaoli-Roach, A. A. and Roach, P. J. (1982) Purification and characterisation of a rabbit liver calmodulin dependent protein kinase able to phosphorylate glycogen synthase. *Journal of Biological Chemistry*, **257**, 8348–55.

Frangakis, M. V., Chatila, T., Wood, E. R. and Sahyoun, N. (1991) Expression of a neuronal Ca²⁺/calmodulin-dependent protein kinase, CaM kinase-Gr, in rat thymus. *Journal of Biological Chemistry*, **266**, 17592–6.

Hanson, P. I. and Schulman, H. (1992) Neuronal Ca²⁺/calmodulin-dependent protein kinases. *Annual Review of Biochemistry*, **61**, 559–601.

Kuret, J. and Schulman, H. (1984) Purification and characterisation of a Ca²⁺/calmodulin dependent protein kinase from rat brain. *Biochemistry*, **23**, 5495–504.

Nairn, A. C. and Greengard, P. (1987) Purification and characterisation of Ca²⁺/calmodulin dependent protein kinase I from bovine brain. *Journal of Biological Chemistry*, **262**, 7273–7281.

Nairn, A. C., Bhagat, B. and Palfrey, H. C. (1985) Identification of calmodulin-dependent protein kinase III and its major Mr 100,000 substrate in mammalian tissues. *Proceedings of the National Academy Sciences USA*, **82**, 7939–43.

Nairn, A. C., Hemmings, H. C. and Greengard, P. (1985) Protein kinases in the brain. *Annual Review of Biochemistry*, **54**, 931–76.

Payne, M. E., Schworer, C. M. and Soderling, T. R. (1983) Purification and characterisation of rabbit liver calmodulin dependent glycogen synthase kinase. *Journal of Biological Chemistry*, **258**, 2376–82.

Woodgett, J. R., Davison, M. T. and Cohen, P. (1983) The calmodulin dependent glycogen synthase kinase from rabbit skeletal muscle: purification, subunit structure and substrate specificity. *European Journal of Biochemistry*, **136**, 481–7.

G Protein-coupled Receptor Kinases

Benovic, J. L., Mayor, F., Staniszeski, C., Lefkowitz, R. J. and Caron, M. G. (1987) Purification and characterisation of the β-adrenergic receptor kinase. *Journal of Biological Chemistry*, **262**, 9026–32.

Debburman, S. K., Ptasienski, J., Boetticher, E., Lomasney, J. W., Benovic, J. L. and Hosey, M. M. (1995) Lipid mediated regulation of G protein coupled receptor kinase 2 and kinase 3. *Journal of Biological Chemistry*, **270**, 5742–7.

Debburman, S. K., Ptasienski, J., Benovic, J. L. and Hosey, M. M. (1996) G protein coupled receptor kinase GRK2 is a phospholipid dependent enzyme that can be conditionally activated by G protein beta/gamma subunits. *Journal of Biological Chemistry*, **271**, 22552–62.

Fredericks, Z. L., Pitcher, J. A., and Lefkowitz, R. J. (1996) Identification of the G protein coupled receptor kinase phosphorylation sites in the human beta(2) adrenergic receptor. *Journal of Biological Chemistry*, **271**, 13796–803.

Kunapuli, P., Onorato, J. J., Hosey, M. M. and Benovic, J. L. (1994) Expression, purification and characterisation of the G protein coupled receptor kinase GRK5. *Journal of Biological Chemistry*, **269**, 1099–105.

Lefkowitz, R. J. (1993) G-protein-coupled receptor kinases. *Cell*, **74**, 409–12.

Palczewski, K. and Benovic, J. L. (1991) G-protein-coupled receptor kinases. *Trends in Biochemical Sciences*, **16**, 387–91.

Premont, R. T., Macrae, A. D., Stoffel, R. H., Chung, N. J. and Pitcher, J. A. (1996) Characterisation of the G protein coupled receptor kinase GRK4: identification of 4 splice varients. *Journal of Biological Chemistry*, **271**, 6403–10.

Winstel, R., Freund, S., Krasel, C., Hoppe, E. and Lohse, M. J. (1996) Protein kinase crosstalk: membrane targetting of the beta adrenergic receptor kinase by protein kinase C. *Proceedings of the National Academy of Sciences USA*, **93**, 2105–9.

Haem-regulated Protein Kinase

Fagard, R. and London, I. M. (1981) Relationship between phosphorylation and activity of heme regulated eukaryotic initiation factor 2-alpha kinase. *Proceedings of the National Academy of Sciences USA*, **78**, 866–70.

Jackson, R. J., Herbert, P., Campbell, E. A. and Hunt, T. (1983) The role of sugar phosphates and thiol reducing systems in the control of reticulocyte protein synthesis. *European Journal of Biochemistry*, **131**, 313–24.

Kozak, M. (1992) Regulation of translation in eukaryotic systems. *Annual Review of Cell Biology*, **8**, 197–225.

Proud, C. G. (1992) Protein phosphorylation in translational control. *Current Topics in Cellular Regulation*, **32**, 243–369.

Samuel, C. E. (1993) The eIF-2 alpha protein kinases, regulators of translation in eukaryotes from yeasts to humans. *Journal of Biological Chemistry*, **268**, 7603–6.

Wettenhall, R. E. H., Kudlicki, W., Kramer, G. and Hardesty, B. (1986) The NH_2 terminal sequence of the alpha subunit and gamma subunits of eukaryote initiation factor II and the phosphorylation site for the heme-regulated EIF-2 alpha kinase. *Journal of Biological Chemistry*, **261**, 2444–7.

Plant-specific Serine/Threonine Kinases

Bennett, J. (1980) Chloroplast phosphoproteins: evidence for a thylakoid-bound phosphoprotein phosphatase. *European Journal of Biochemistry*, **104**, 85–9.

Coughlan, S. J. and Hind, G. (1986) Purification and characterisation of a membrane-bound protein kinase from spinach thylakoids. *Journal of Biological Chemistry*, **261**, 11378–85.

Ranjeva, R. and Boudet, A. M. (1987) Phosphorylation of proteins in plants: regulatory effects and potential involvement in stimulus response coupling. *Annual Review of Plant Physiology*, **38**, 73–93.

Stone, J. M. and Walker, J. C. (1995) Plant protein kinase families and signal transduction. *Plant Physiology*, **108**, 451–7.

Tyrosine Kinases

Bellot, F., Crumley, G., Kaplow, J. M., Schessinger, J., Jaye, M. and Dionne, C. A. (1991) Ligand-induced transphosphorylation between different FGF receptors. *EMBO Journal*, **10**, 2849–54.

Blackshear, P. J., Hampt, D. M., App, H. and Rapp, U. R. (1990) Insulin activates the Raf-1 protein kinase. *Journal of Biological Chemistry*, **265**, 12131–4.

Cooper, J. A., Esch, F. S., Taylor, S. S. and Hunter, T. (1984) Phosphorylation sites in enolase and lactate dehydrogenase utilised by tyrosine kinases *in vivo* and *in vitro*. *Journal of Biological Chemistry*, **259**, 7835–41.

Duan, D.-S. R., Pazin, M. J., Fretto, L. J. and Williams, L. T. (1991) A functional soluble extracellular region of the platelet derived growth factor (PDGF) β–receptor antagonises PDGF-stimulated responses. *Journal of Biological Chemistry*, **266**, 413–18.

Fantl, W. J., Escobedo, J. A., Martin, G. A.*et al.* (1992) Distinct phosphotyrosines on a growth factor receptor bind to specific molecules that mediate different signalling pathways. *Cell*, **69**, 413–23.

Fantl, W. J., Johnson, D. E. and Williams, L. T. (1993) Signalling by receptor tyrosine kinases. *Annual Review of Biochemistry*, **62**, 453–81.

Kashles, O., Yarden, Y., Fischer, R., Ullrich, A. and Schlessinger, J. (1991) A dominant negative mutation suppresses the function of normal epidermal growth factor receptors by heterodimerisation. *Molecular and Cellular Biology*, **11**, 1454–63.

Kawamura, M., McVicar, D. W., Johnston, J. A. *et al.* (1994) Molecular cloning of L-Jak, a Janus family protein-tyrosine kinase expressed in natural killer cells and activated leukocytes. *Proceedings of the National Academy of Science USA*, **91**, 6374–8. Known now as JAK3.

Kovacina, K. S., Yonezawa, K., Brautigan, D. L., Tonks, N. K., Rapp, U. R. and Roth, R. A. (1990) Insulin activates the kinase activity of the Raf-1 proto-oncogene by increasing its serine phosphorylation. *Journal of Biological Chemistry*, **265**, 12115–18.

Schlessinger, J and Ullrich, A. (1992) Growth factor signalling by receptor tyrosine kinases. *Neuron*, **9**, 383–91.

Siew Lai, K., Jin, Y., Graham, K. D., Witthuhn. B. A., Ihle, J. N. and Liu, E. T. (1995) A kinase deficient splice variant of the human JAK3 is expressed in hematopoietic and epithelial cancer cells. *Journal of Biological Chemistry*, **270**, 25028–36.

Ueno, H., Colbert, H. A., Escobedo, J. A. and Williams, L. T. (1991) Inhibition of PDGFβ receptor signal transduction by coexpression of a truncated receptor. *Science*, **252**, 844–8.

Mitogen-activated Protein Kinases

Blenis, J. (1993) Signal transduction via the MAP kinases: proceed at your own RSK. *Proceedings of the National Academy of Sciences USA*, **90**, 5889–92.

Blumer, K. J. and Johnson, G. L. (1994) Diversity in function and regulation of

MAP kinase pathways. *Trends in Biochemical Sciences*, **19**, 236–40.

Boulton T. G., Nye, S. H., Robbins, D. J. *et al.* (1991) ERKs: a family of protein serine/thronine kinases that are activated and tyrosine phosphorylated in response to insulin and NGF. *Cell*, **65**, 663–75.

Fialkow, L., Chan, C. K., Rotin, D., Grinstein, S. and Downey, G. P. (1994) Activation of the mitogen-activated protein kinase signalling pathway in neutrophils. *Journal of Biological Chemistry*, **269**, 31234–42.

Herskowitz, I. (1995) MAP kinase pathways in yeast: for mating and more. *Cell*, **80**, 187–97.

Irie, K., Takase, M., Lee, K. S. *et al.* (1993) *MKK1* and *MKK2*, which encode *Saccharomyces cerevisiae* mitogen-activated protein kinase-kinase homologs, function in the pathway mediated by protein kinase C. *Molecular and Cellular Biology*, **13**, 3076–83.

Marshall, C. J. (1994) MAP kinase kinase kinase, MAP kinase kinase and MAP kinase. *Current Opinion in Genetics and Development*, **4**, 82–9.

Nebreda, A. R. (1994) Inactivation of MAP kinases. *Trends in Biochemical Sciences*, **19**, 1–2.

Pelech S. L. and Sanghera, J. S. (1992) MAP kinases: charting the regulatory pathway. *Science*, **257**, 1355–6.

Roberts, R. L. and Fink, G. R. (1994) Elements of a single MAP kinase cascade in *Saccharomyces cerevisiae* mediate two developmental programs in the same cell type: mating and invasive growth. *Genes and Development*, **8**, 2974–85.

Stevenson, M. A., Pollock, S. S., Coleman, C. N. and Calderwood, S. K. (1994) X-irradiation, phorbol esters, and H_2O_2 stimulate mitogen-activated protein kinase activity in NIH-3T3 cells through the formation of reactive oxygen intermediates. *Cancer Research*, **54**, 12–15.

Thomas, G. (1992) MAP kinase by any other name smells just as sweet. *Cell*, **68**, 3–6.

Histidine Phosphorylation

Alex, L. A. and Simon, M. L. (1994) Protein histidine kinases and signal transduction in prokaryotes and eukaryotes. *Trends in Genetics*, **10**, 133–8.

Borkovich, K. A., Kaplan, N., Hess, J. F. and Simon, M. I. (1989) Transmembrane signal transduction in bacterial chemotaxis involves ligand-dependent activation of phosphate group transfer. *Proceedings of the National Academy of Sciences USA*, **86**, 1208–12.

Kofoid, E. C. and Parkinson, J. S. (1992) Communication modules in bacterial signalling proteins. *Annual Review of Genetics*, **26**, 71–112.

Stock, J., Mottenen, J. M., Stock, J. B. and Schutt, C. E. (1989) Three dimensional structure of CheY, the response regulator of bacterial chemotaxis. *Nature*, **337**, 745–9.

Volz, K. and Matsumura, P. (1991) Crystal structure of *Escherichia coli* CheY refined at 1.7Å resolution. *Journal of Biological Chemistry*, **266**, 15511–19.

Phosphatases

Cohen, P. (1989) The structure and regulation of protein phosphatases. *Annual Review of Biochemistry*, **58**, 453–508.

Cohen, P. and Cohen, P. T. W. (1989) Protein phosphatases come of age. *Journal of Biological Chemistry*, **264**, 21435–8.

Cohen, P. T. W., Brewis, N. D., Hughes, V. and Mann, D. J. (1990) Protein serine/threonine phosphatases: a expanding family. *FEBS Letters*, **268**, 355–9.

Fischer, E. H., Charbonneau, H. and Tonks, N. K. (1991) Protein tyrosine phosphatases: a diverse family of intracellular and transmembrane enzymes. *Science*, **253**, 401–6.

Frangioni, J. V., Beahm, P. H., Shifrin, V., Jost, C. A. and Neel, B. G. (1992) The nontransmembrane tyrosine phosphatase PTP-1B localises to the endoplasmic reticulum via its 35 amino acid C-terminal sequence. *Cell*, **68**, 545–60.

Guan, K. L., Broyles, S. S. and Dixon, J. E. (1991) A Tyr/Ser protein phosphatase encoded by vaccinia virus. *Nature*, **350**, 359–62.

Hunter, T. and Sefton, B. M. (1980) Transforming gene product of Rous sarcoma virus phosphorylates tyrosine. *Proceedings of the National Academy of Sciences USA*, **77**, 1311–15.

Pot, D. A., Woodford, T. A., Remboutsika, E., Haun, R. S. and Dixon, J. E. (1991) Cloning, bacterial expression, purification and characterisation of the cytoplasmic domain of rat LAR, a receptor-like protein tyrosine phosphatase. *Journal of Biological Chemistry*, **266**, 19688–96.

Tonks, N. K., Charbonneau, H., Diltz, C. D., Fischer, E. H. and Walsh, K. A. (1988) Demonstration that the leukocyte common antigen CD45 is a protein tyrosine phosphatase. *Biochemistry*, **27**, 8695–701.

Tonks, N. K., Diltz, C. D. and Fischer, E. H. (1988) Characterisation of the major protein-tyrosine phosphatases of human placenta. *Journal of Biological Chemistry*, **263**, 6731–7

Walton, K. M. and Dixon, J. E. (1993) Protein tyrosine phosphatases. *Annual Review of Biochemistry*, **62**, 101–20.

cAMP, Adenylyl Cyclase and the Role of G Proteins

Introduction

Once the cell has received an external signal, such as the binding of a ligand to a cell-surface receptor, a second-messenger response inside the cell rapidly follows. One of the major second messengers is the intracellular concentration of cyclic adenosine monophosphate or adenosine $3',5'$-cyclic monophosphate to give it its full chemical name. It is understandably simply referred to as cyclic AMP or more commonly cAMP. Some of the hormones that have their cellular effects mediated by cAMP include epinephrine and glucagon along with those listed in Table 5.1.

Table 5.1. Some signalling molecules that act via cAMP.

Hormone	Major tissues affected
Adrenocorticotropic hormone	Adrenal cortex, fat
Epinephrine (adrenaline)	Muscle, fat, heart
Glucagon	Liver, fat
Luteinising	Ovaries
Parathormone	Bone
Thyroid-stimulating hormone	Thyroid gland, fat
Vasopressin	Kidney

cAMP is produced at the plasma membrane of the cell by the enzyme complex adenylyl cyclase, alternatively known as adenylate cyclase, from which the cAMP is released into the cytosol where it can diffuse and act on the next part of the signal-transduction pathway. For example, the next response might be the binding of cAMP to cAPK, so increasing the phosphorylation of certain proteins with concomitant alteration of their activity. An excellent example of this is the control of glycogen metabolism, as illustrated in Figure 5.1. An extracellular signal such as epinephrine

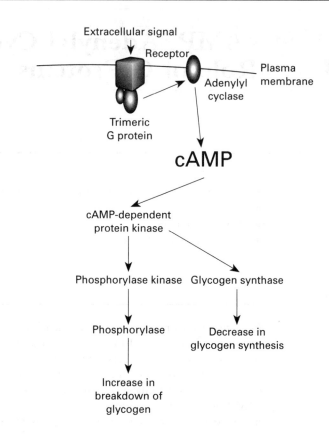

Figure 5.1
Coordinated control of glycogen metabolism. Hormones binding to receptors activate a G protein (G$_s$), which activates adenylyl cyclase. The rise in cAMP activates cAMP-dependent protein kinase, leading to the phosphorylation of phosphorylase kinase and glycogen synthase, causing inhibition of the latter. Phosphorylation of phosphorylase by phosphorylase kinase causes an increase in glycogen breakdown. Therefore the overall result is an inhibition of glycogen synthesis and in increase in glycogen breakdown.

binds to a membrane receptor, which in turn activates a G protein. The G protein activates adenylyl cyclase to produce cAMP. The increase in intracellular cAMP subsequently activates cAPK, which phosphorylates phosphorylase kinase. This latter enzyme phosphorylates, and in so doing activates, phosphorylase, which breaks down glycogen, releasing glucose for energy metabolism. cAPK also phosphorylates and inhibits glycogen synthase, reducing the synthesis of glycogen in the same cell.

This route, like most signalling pathways, leads to massive amplification of the signal. The activation of one receptor leads to the activation of several adenylyl cyclase enzymes, each leading to the production of hundreds or thousands of cAMP molecules. cAMP is ideal as a second-messenger molecule because it is rapidly made, small and readily diffusible and is readily and quickly broken down by another enzyme, phosphodiesterase. Therefore the signal can be rapid and reversible.

As well as the activation of cAPK, cAMP also has more direct effects. For example, in *E. coli* it acts through the *crp* gene product, which is known as cAMP receptor protein (CRP, alternatively known as CAP). This is a dimer of 22 kDa subunits. An increase in intracellular cAMP concentration causes cAMP to bind to CRP to form a cAMP–CRP (or cAMP–CAP) complex, which has been shown to bind DNA and alter gene expression of several genes, including those of the lactose and arabinose operons.

However, cAMP can also be an extracellular signal, acting as a type of hormone in the slime mould *Dictyostelium*, where cAMP is used as a signal between cells, controlling cellular aggregation and differentiation.

Adenylyl Cyclase

In mammals, adenylyl cyclase is a single polypeptide that resides in the plasma membranes of cells. Its role is to catalyse production of cAMP from ATP. The 3'-OH ribose group of the ATP attacks the α-phosphoryl group resulting in cyclisation of the molecule. The by-product from the reaction is inorganic pyrophosphate, which is itself broken down by an enzyme, pyrophosphatase. The energy released from this latter reaction helps to drive the cyclisation reaction (Figure 5.2).

Figure 5.2
Production and hydrolysis of cAMP.

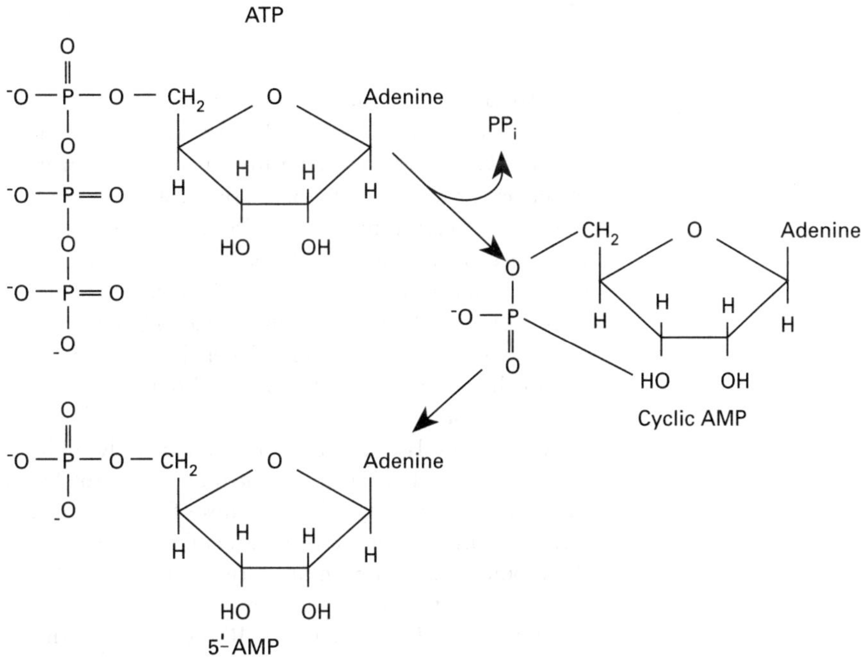

Like many of the proteins in signal transduction, there are various isoforms of adenylyl cyclase. Although the mammalian ones are integral in the plasma membrane, other forms have been found to be peripheral plasma-membrane proteins, for example in the yeast *S. cerevisiae* or in *E. coli.*; even soluble forms have been reported in some bacteria. A soluble form has also been suggested to exist in mammalian sperm. At least eight isoforms of the plasma-membrane adenylyl cyclase have been identified in mammals (types I–VIII), but all share similar structural topology. In general, they contain two clusters of six transmembrane-spanning highly hydrophobic domains that separate two catalytic domains on the cytoplasmic side of the membrane (Figure 5.3). Areas of

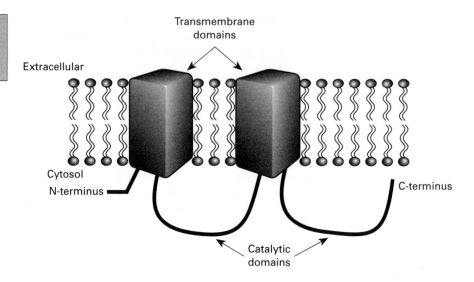

Figure 5.3
Proposed structural topology of adenylyl cyclase.

the two catalytic domains are well conserved and also homologous to regions of membrane-bound guanylyl cyclases. It is probable that both these conserved regions, called C_{1a} and C_{2a}, are responsible for the activity of the enzyme, as point mutations in either region are extremely inhibitory. Using a study of the hydrodynamic properties of the enzyme, a molecular mass of 200-250 kDa has been proposed; however, the actual molecular mass is approximately 120 kDa, begging the question as to whether the enzyme exists as a dimer in the membrane. The extra apparent molecular mass may come from its association with other proteins, such as G proteins or even receptors.

Although one of the forms of adenylyl cyclase from *Dictyostelium* conforms to the mammalian model, another form differs in only having one transmembrane-spanning region.

Adenylyl cyclases can be non-competitively inhibited by Mg^{2+}/ATP analogues such as 3'-AMP but can be activated by a lipid-soluble diterpene, forskolin. It is presumed that the hydrophobic domains of the enzyme are the target for forskolin. In fact, isoforms without the six transmembrane spans seem to be insensitive to forskolin. A hypothesis has also been put forward that the hydrophobic domains might act as a channel across the membrane, and this is seen as important in cells that excrete cAMP. For example, in *Dictyostelium*, cAMP is used as a signal for cellular aggregation and differentiation. However, no concrete evidence for adenylyl cyclase acting as a transporter has been found and the heterogeneity of the sequences in these regions between different isoforms would suggest that this is not a role of the enzyme.

Adenylyl Cyclase Control and the Role of G Proteins

When a hormone binds to the relevant receptor on the cell surface, activation of adenylyl cyclase is not a direct process, but instead, requires the use of other proteins; further, it was discovered that breakdown of GTP was also involved. The other proteins are in fact guanyl nucleotide-binding proteins or G proteins. Activation of the receptor in turn causes the activation of the G protein, which regulates the activity of adenylyl cyclase. Some extracellular signalling molecules that bind to G protein-linked receptors are discussed in Chapter 3 (see Table 3.1).

The action of a G protein on the activity of adenylyl cylcase is either stimulatory or inhibitory, depending on the G protein involved. If the result of G-protein activation is stimulation of adenylyl cyclase activity, the G protein is known as stimulatory G protein or G_s. In the inactive state this G protein exists as a complex of three polypeptides, α, β and γ: α has a molecular mass of approximately 45 kDa, β of approximately 35 kDa and γ of only 7 kDa. Hence these G proteins are known as the trimeric G proteins.

In the inactive state the G protein has GDP bound to a single GDP-binding site. On activation of the receptor, the GDP on the G protein is released in exchange for GTP. This causes the breakdown of the G protein complex into a free α subunit bound to GTP and a β/γ complex that does not dissociate further. The free α subunit–GTP diffuses to adenylyl cyclase, where it binds and causes activation, resulting in the subsequent release of cAMP (Figure 5.4). Any signal that is turned on needs

Figure 5.4
Regulation of adenylyl cyclase by the activation of a trimeric G protein by a G protein-linked receptor.

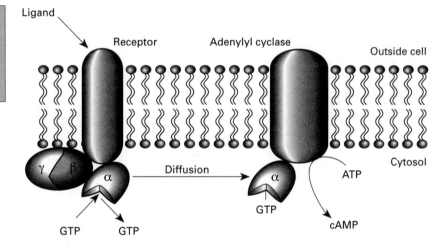

to be turned off again. Deactivation of the G protein is achieved by the breakdown of GTP to GDP on the α subunit. The breakdown of GTP is catalysed by the intrinsic GTPase activity of the α subunit itself. This GTPase activity is activated by the binding of the α subunit to adenylyl cyclase, with the α subunit subsequently dissociating from the adenylyl cyclase and reforming the original complex with the β/γ subunits (Figure 5.5).

Certain oncogenes have been shown to code for G proteins that contain a defect in their intrinsic GTPase activity. Although not part of the trimeric family of G proteins, the oncogene product Ras is a good example. Such proteins once bound to GTP are unable to return to the inactive state, causing the cell to continue to receive the 'on' signal, even in the absence of receptor binding to a ligand (Figure 5.6).

Figure 5.5
G protein cycle: activation by nucleotide exchange and subunit dissociation; deactivation by nucleotide hydrolysis and subunit reassociation.

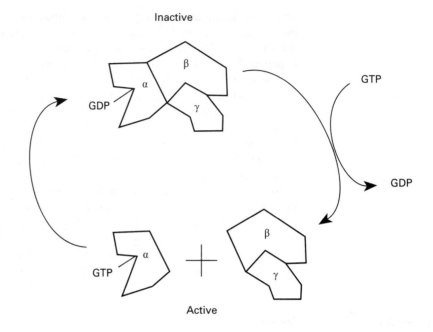

Figure 5.6
Blockade of the G protein cycle causes permanent activation of G protein.

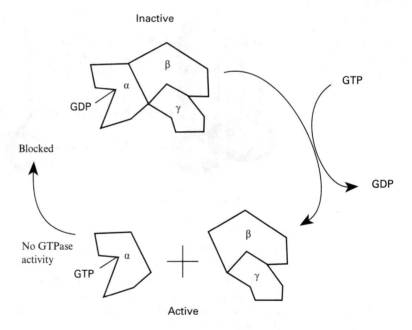

A similar effect can be emulated *in vitro* by the addition of GTPγ-S. This form of GTP cannot be broken down by GTPase and again the α subunit remains in an active state. The importance of turning off the G protein signal is emphasised by the disease cholera. The toxin released from the bacterium *Vibrio cholerae* inactivates the α subunit of these G proteins. Cholera toxin, otherwise known as choleragen, is in fact an oligomeric enzyme, consisting of an A_1 subunit (25 kDa) linked to an A_2 subunit (5.5 kDa) by a disulphide bond along with five B subunits (each 16 kDa). The A_1 subunit of the enzyme uses NAD^+ as a substrate, catalysing the transfer of ADP-ribose to a specific arginine residue of the α subunit, destroying the α subunit's intrinsic GTPase activity. The efficient modification of $G_{s\alpha}$ requires the presence of another G protein called ADP-ribosylation factor (ARF), which is a subgroup of the Ras family of proteins. Once modified, $G_{s\alpha}$ permanently activates adenylyl cyclase and the subsequent loss of cAMP control in the gut epithelial cells leads to the movement of Na^+ and water into the intestine, causing the resultant diarrhoea. The $G_{s\alpha}$ subunit can also be activated by the presence of aluminium tetrafluoride (AlF_4^-) together with Mg^{2+}.

It needs to be noted that the G_s protein is not a single protein but a family of G protein subunits, with $G_{s\alpha}$ arising from alternative splicing of the same mRNA. This leads to at least four forms of the $G_{s\alpha}$ subunit.

G proteins are not only able to cause activation of adenylyl cyclase but another form can lead to inhibition of the enzyme. This G protein, termed G_i, is analogous to G_s described above, except that on activation and diffusion to adenylyl cyclase the release of cAMP is depressed. The G_i protein contains the same β/γ subunit complex as G_s, but the G_α subunit differs. There are two possible ways in which this G_i protein causes the inhibition of adenylyl cyclase. Firstly, the $G_{i\alpha}$ subunit once released may be able to cause direct or indirect inhibition of adenylyl cyclase. Secondly, the $G_{i\alpha}$ subunit is much less inhibitory than the β/γ subunit released by the G protein's activation. Interestingly, the β/γ subunit is much more inhibitory in the presence of G_s, suggesting that the mode of action of the β/γ subunit from G_i is in fact to bind to $G_{s\alpha}$ from the dissociation of G_s and stop activation of adenylyl cyclase by the G_s route. This would be possible as the β/γ complex of G_i and G_s are the same. This interaction of the β/γ complex with different G_α subunits has been termed the subunit exchange (Figure 5.7). In a similar way to the permanent activa-

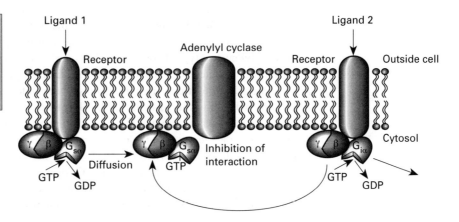

Figure 5.7
Subunit exchange: activation of G_i prevents stimulation by G_s through the action of the β/γ subunits.

tion of $G_{s\alpha}$ by cholera toxin, the $G_{i\alpha}$ subunit is susceptible to modification mediated by another toxin, this time pertussis toxin. This is released by the bacterium that causes pertussis or whooping cough. In this case, the ADP-ribose moiety is added to a specific cysteine residue at the C-terminal end of the $G_{i\alpha}$ subunit, destroying its interaction with the receptor and so preventing its activation. The lack of activation of the G_i protein stops the inhibition of adenylyl cyclase.

It should be noted, however, that several of these membrane-associated G proteins are not activated or inhibited either by cholera toxin or pertussis toxin. Amongst these, a subfamily known as G_q has been reported to be important in the control of PLC activity.

The interplay of G_s and G_i proteins is illustrated by the action of epinephrine on different receptors. On binding to the cell's β-adrenergic receptors, the activity of adenylyl cyclase is increased through a route that uses G_s. On the other hand, epinephrine binding to α_2-adrenergic receptors causes a decrease in the activity of the cell's adenylyl cyclase through a route that uses G_i.

The exact nature of the binding of adenylyl cyclase to G proteins has been investigated by the use of chimeric G proteins, i.e., proteins composed of different parts of different G proteins. For example, part of the α chain may be derived from an α subunit known not to stimulate adenylyl cyclase, such as a chimera made of segments from G_{i2} and G_t. G_t is a G protein known as transducin and has a defined role in photoreception of the rod cells of the eye (discussed in Chapter 9). By the use of various combinations in such chimeras, the active parts of the α subunits can be deciphered. For example, if the protein contained 40% of the C-terminal end of the G_s subunit, adenylyl cyclase was activated, indicating that this was the crucial part of the polypeptide for this interaction.

Although the active part of trimeric G proteins has traditionally been thought to be the α subunit, the β/γ complex has been shown to be also important. Like the studies with chimeric α subunits, as many isoforms of the β and γ subunits exist, defined β/γ complexes with known β and γ subunits were formed by expression of the subunits in insect Sf9 cells. These could be subsequently purified by subunit affinity chromatography and used to study if there was any interaction of the β/γ complexes with adenylyl cyclase. It transpired that different isoforms of adenylyl cyclase are controlled differently by the subunits of the G protein trimer. All the mammalian enzymes are activated by G_s, but the action of β/γ can be stimulatory, as with adenylyl cyclase types II or IV, or inhibitory, as seen with adenylyl cyclase type I. Furthermore, some adenylyl cyclase isoforms do not appear to be affected by β/γ at all. It was also found that the interactions of the different isoforms were dependent on prenylation of the γ subunit of the G protein.

Interestingly, it was also noted that the concentration of β/γ needed to regulate adenylyl cyclase is greater than the concentration of $G_{s\alpha}$ released; therefore, as it is a stoichiometric dissociation when the G protein is activated, the β/γ probably does not arise solely from the breakdown of the G_s trimer. The β/γ complex may come from G_i or another G protein, G_o, which are more abundant in some tissues, such as the brain. This sort of interaction may be commonplace, enabling different G protein trimers to have different effects on adenylyl cyclase at the same time.

Besides the possible regulation of adenylyl cyclase, other roles in which the β/γ subunits are involved include the regulation of K^+ channels in myocytes and the pheromone signalling pathway of yeasts, which involves the activation of MAPK cascades. The β/γ complex has also been implicated in the control of the kinase activity and phosphorylation of β-adrenergic receptors. It may that β/γ interacts either with the kinase or the receptor. Another interaction reported is with a protein called phosducin, which may be involved in regulation of the concentration of β/γ by forming a complex with it. In the photoreceptor cells of the eye, phosducin is a phosphoprotein whose phosphorylated state is light sensitive, mediated by light-induced changes in cyclic nucleotide concentrations. In the dark, the protein is phosphorylated by cAPK, while it is dephosphorylated in the light by phosphatase 2A. The role of phosducin seems to be to bind with the β/γ complex thereby stopping its interaction with the G_α subunit and preventing the G_α subunit from recycling. Similar proteins to phosducin have been reported to be present in the cytosol of bovine brain cells, and this type of control might be quite ubiquitous.

The β/γ subunit has also been shown to activate one of the forms of PLC, i.e PLCβ, in the promyeloid HL60 cell, leading to activation of the inositol phosphate pathway. The isoforms of PLC that appear to be most susceptible are β2 and β3.

The α subunit released from the activation of G_i, i.e. the $G_{i\alpha}$ subunit, is also able to exert an effect on enzymes other than adenylyl cyclase. In muscle tissue $G_{i\alpha}$ is able to activate K^+ membrane channels, allowing K^+ to exit the cells thus inhibiting contraction of the muscle. Likewise, the $G_{s\alpha}$ subunit inhibits cardiac Na^+ channels and an effect of $G_{s\alpha}$ on voltage-gated Ca^{2+} channels has also been reported. Thus the exact signalling process that transpires from dissociation of the G protein trimer may be very complex.

If these subunits from G proteins are readily diffusible along the membrane from receptor to enzyme, what is it that keeps them attached to the membrane when dissociation takes place and stops them from diffusing out into the cytosol? For the γ subunit it has been proposed that the subunit is bound to an isoprenoid lipid, which being a hydrophobic molecule attaches the subunit to the membrane. For the α subunits, the attachment is probably through another lipid, this time myristic acid.

Although classically these G proteins have been thought of as activators of enzymes on the plasma membrane, particularly adenylyl cyclase, it is thought that their action may be far more widespread throughout the cell. For example, an isoform of the subunit $G_{i\alpha}$ has been localised to the Golgi apparatus, where it is probably involved in the control of the packaging of proteins and in the regulation of vesicular activity.

It is likely that the role of G proteins in disease and disorders will be far more widespread than reported to date. Defects of the G_s and G_i proteins have been implicated in the excessive proliferation of pituitary cells leading to pituitary cancer. It is reasonably certain that the role of these G proteins will be implicated in many disorders and be the target for many therapies.

However, it should be mentioned that G proteins are not the only means of controlling the activity of adenylyl cyclase. Ca^{2+}/calmodulin can also activate some of the isoforms of this enzyme, particularly type I but also types III and VIII as well as some non-mammalian forms. Types V and VI seem to show inhibition by Ca^{2+}, while types II and VII are stimulated by PKC.

Although the discussion above has concentrated on G_s and G_i, since the discovery of trimeric G proteins in the early 1980s, at least 20 different forms have been isolated and the number of different receptors that have their effects through the action of G proteins must run into the hundreds. The different α subunits of the trimer are used in general to define the G protein, but as well as the different α subunits at least five iso-forms of the β subunit exist and more than 10 γ subunits have been identified. If every type of α, β and γ subunit could randomly mix in the trimer then there could be several hundreds of different forms of the G protein. However, reconstitution studies show that not all combinations are possible. For example, $G_{\gamma1}$ can complex with $G_{\beta1}$ but not with $G_{\beta2}$. It is probable, in fact, that the stability of some of the subunits may be dependent on the existence of, and binding with, other relevant partners. If G_β is formed in the absence of a G_γ able to complex with it, then it is probably aggregated and degraded. Many combinations expressed will be tissue specific.

Other trimeric G proteins that have been characterised include G_o of brain neurones, G_{olf} and G_t of the olfactory cells and light-sensitive cells of the eye respectively and G_q of smooth muscle cells of the blood vessels. These do not all have an effect on adenylyl cyclase, for example G_q regulates the action of PLC, which leads to the release of other intracellular messengers through the inositol phosphate pathway, leading to the release of Ca^{2+} and the activation of phosphorylation, while G_t controls the activity of cGMP phosphodiesterase. Some of these G proteins are listed in Table 5.2, while the functions of G proteins are considered again in Chapter 10.

Table 5.2. The more common members of the G protein α subunit families.

G_s

α_{olf}	α_s			

G_i

α_{i1}	α_{i2}	α_{i3}		
α_{oA}	α_{oB}			
α_{t1}	α_{t2}			
α_g				
α_z				

G_q

α_q	α_{11}	α_{14}	α_{15}	α_{16}

G_{12}

α_{12}	α_{13}

The ability to stimulate trimeric G proteins artificially in the laboratory will help greatly in the elucidation of their role in new pathways and systems. For example, a tetradecapeptide isolated from wasp venom, mastoparan, mimicks the action of G protein-coupled receptors and causes the activation of G proteins, particularly G_i and G_o.

Contrary to the above discussion where the activation of G proteins leads to control of the production of cAMP, control may be in the reverse direction where the levels of cAMP regulate activation of the G protein. In the slime mould *Dictyostelium* the level of extracellular cAMP is detected by receptors that lead to the activation of the G protein subunit $G_{\alpha2}$.

In an analogous way to the production of cAMP, cGMP is also used by cells as a signalling molecule. cGMP was identified from urine in 1963 by Ashman and his colleagues and is produced by the enzyme guanylyl cyclase, otherwise known as guanylate cyclase. Unlike adenylyl cyclase, the cAMP-producing enzyme, guanylyl cyclase is found in two forms, a soluble type that resides in the cytoplasm of the cell and a membrane-bound form located in the plasma membrane. These forms are distinct proteins with their own modes of regulation.

The classical role for cGMP is in the regulation of cGMP-gated ion channels in the photosensitive cells of the retina (discussed further in Chapter 10). However, cGMP has also been found to control cGMP-dependent phosphodiesterases and cGPKs. It has also been suggested that the hydrolysis of cGMP and the subsequent energy release is the important factor in signalling by cGMP rather than the measurable rise in the concentration of this molecule.

Soluble Guanylyl Cyclase

The soluble form of guanylyl cyclase is characterised by the presence of haem at the catalytic site. The haem prosthetic group is an iron-containing protoporphyrin IX, as it is in most cytochrome molecules in cells. In fact protoporphyrin IX acts as a strong activator of the enzyme, while it is the haem group that is the site of regulation by nitric oxide (NO)(discussed in Chapter 8).

The topology of the enzyme shows it to be a heterodimer of α and β subunits of approximately 70 kDa and 85 kDa, with each subunit containing a region homologous to the C_{1a} and C_{2a} regions of adenylyl cyclase; both are subunits required for catalysis. The presence of Mn^{2+} ions as opposed to Mg^{2+} increases the activity of the enzyme, which also is slightly stimulated by Ca^{2+} ions. However, ATP acts as an inhibitor. Not surprisingly haem binding has been shown to involve a histidine residue, which is on the β subunit of the enzyme.

However, the enzyme appears to exist in tissue-specific isoforms. Two isoforms have been reported to exist in bovine lung: the two forms contain one subunit in common but the second subunit differs, one being 85 kDa while the other is only 73 kDa. Interestingly, the haem spectrum for the enzyme differs slightly depending upon the tissue from which the enzyme has been purified, and this suggests that the attachment of the haem group may vary. It has been suggested that the haem group of the bovine lung enzyme is penta-coordinate while that of the human placenta is hexa-coordinate, which implies that the two enzymes may have subtle differences in their catalytic action and in their control.

Membrane-bound Guanylyl Cyclase

Besides the soluble forms of guanylyl cyclase, there are also membrane-bound forms. These contain a single membrane-spanning domain, along with a protein-kinase homology domain on the cytoplasmic side of the membrane and an intracellular catalytic domain (Figure 5.8). Some areas of homology to the mammalian adenylyl cyclases have been noted. These membrane forms of guanylyl cyclase also have an extracellular domain and act as receptors for two main classes of molecules: signalling peptides, or the so-called natriuretic peptides, and the heat-stable enterotoxins including guanylins.

Figure 5.8
Domain structure of membrane-associated guanylyl cyclase.

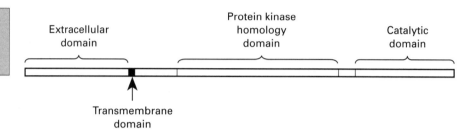

These membrane cyclase forms, like many receptor proteins, probably function as dimers or oligomers. ATP has been shown to stimulate the activity of these cyclases but, despite the presence of a protein-kinase homology domain, non-hydrolysable forms of ATP have the same effect and neither ATPase activity nor phosphorylation activity is required for the functioning of these receptors. However, it appears that the cyclase normally resides in a highly phosphorylated state. Further, a novel protein phosphatase has been found associated with one form of the receptor and it is thought that the phosphorylation state of the cyclase may be crucial to its activity, dephosphorylation causing desensitisation of the receptor.

Ca^{2+} ion concentrations are also important in the regulation of the membrane forms of guanylyl cyclase. In the retina, Ca^{2+}-sensitive proteins have been found that interact with guanylyl cyclase. One such protein is recoverin, a 26 kDa protein. Other such proteins have molecular mass of 20 kDa and 24 kDa, known as p20 and p24 respectively. These proteins are located in the cytoplasm, where they respond to changes in the Ca^{2+} concentration, but their exact interaction with the cyclase remains to be resolved. However, increases in Ca^{2+} lead to a complex of the Ca^{2+} sensitive protein and the cyclase and the production of cGMP is decreased. It is possible that Ca^{2+} might also have a direct interaction with the cyclase but this has never been demonstrated.

Interestingly, one particulate form from bovine rod outer segments is reported to be activated up to 20-fold by NO, suggesting the involvement of a haem group, analogous to that found with the soluble forms of guanylyl cyclase.

Phosphodiesterases

The breakdown of cAMP and cGMP and hence the reduction of their intracellular concentrations, which leads to cessation or indeed propagation of the signal, is catalysed by the enzymes called cyclic nucleotide phosphodiesterases (PDEs), usually referred to simply as phosphodiesterases. This reaction can be written as follows (see also Figure 5.2):

$$cAMP + H_2O \xrightarrow{Mg^{2+}} AMP + H^+$$

Although cAMP is relatively stable in the cell, removal of the cyclic nature of the molecule and hence its breakdown is overall exergonic, i.e. the change in Gibbs free energy (ΔG) is negative, the enzyme overcoming the activation energy barrier that stops the reaction becoming spontaneous.

In mammals, at least seven different types of PDE have been distinguished (Table 5.3), with an eighth class suggested by some. As would be expected some of the types

Table 5.3. Classes of phosphodiesterases.

Class of enzyme	Characteristics
I	Ca^{2+}/calmodulin dependent
II	cGMP stimulated
III	cGMP inhibited
IV	cAMP specific
V	cGMP specific
VI	Photoreceptor type
VII	High-affinity cAMP specific

have more than one form, for example type III has at least two forms, cGIP1 and cGIP2, which are highly homologous in their C-terminal regions but show greater divergence towards the N-terminal end of the polypeptides. The expression of such different forms of the same family of PDEs is also tissue specific, suggesting subtle differences in their functionality. If the cDNAs for all the PDEs are compared, the 3'ends show a great deal of homology over a stretch of approximately 800 base pairs. These regions code for the common catalytic core of the molecules. However, comparison of the 5' ends shows low homology and accounts for the differences that make up the seven family types. The N-terminal end of the polypeptides may also influence their subcellular location, either containing signal peptides that direct their post-translational route within the cell or allowing their attachment to membranes.

The genes for five cAMP-specific PDEs have been located in both humans and mice. In humans, two of the genes are found on chromosome 19, with others on chromosomes 1, 5 and 8; in mice, PDE genes are found on chromosomes 4, 8, 9 and 13. However multiple forms of a PDE within a family may also arise from the same gene. In some cases multiple promoters within one gene have been found, enabling transcription to start at different points and therefore resulting in polypeptides of varying length. Even if the same promoter is used, the gene transcript may undergo alternative splicing and varying polypeptides again result. Such differences in expression and post-transcriptional modification may be tissue specific, each of the PDEs retaining the catalytic core but having alternative modules that alter their function and cellular location. However, some PDEs are particularly susceptible to proteolysis and thus several reports in the literature of the presence of new isoenzyme forms may only be the result of breakdown products from larger PDEs.

Although cAMP and cGMP are often broken down by different PDEs, the presence of one cyclic nucleotide may still influence the intracellular concentration of the other. Some forms of cAMP PDE are influenced by the presence of cGMP. For example, type III has been shown to be inhibited by the presence of cGMP. Such PDEs are therefore referred to as cGMP-inhibited phosphodiesterases (cGI-PDEs). These PDEs have a molecular mass of approximately 110 kDa, although proteolytic products of 30-80 kDa have been reported also. Other cAMP PDEs contain allosteric binding sites for cGMP that cause stimulation. These PDEs have an apparent molecular mass of approximately 105 kDa; further analysis suggests that the native structure consists of non-spherical dimers, although a tetrameric form has been isolated from rabbit brain tissue.

It is also possible that the binding of cyclic nucleotides to PDEs takes place without catalytic turnover. This might serve to buffer the intracellular concentration of such signalling molecules, i.e. a cyclic nucleotide such as cGMP is not necessarily free in solution to have a further effect on other enzymes.

Another family of PDEs are controlled by the intracellular concentration of Ca^{2+} in association with calmodulin. This family has been further subdivided into four groups, which have the molecular mass of 60, 63, 58 and 67 kDa.

The most well-characterised cGMP PDE is that from the rod cells of the eye. As discussed in Chapter 10, cGMP acts as a regulator of ion channels. The membrane-bound PDE from bovine rods exists as an $\alpha\beta\gamma_2$ complex, where the α subunit has a molecular mass of 88 kDa, the β subunit has a molecular mass of 84 kDa while the γ subunit has an apparent molecular mass of only 11 kDa. However, other complexes may exist such as $\alpha\alpha\gamma_2$ and $\beta\beta\gamma_2$. As well as the membrane form, there is also a soluble form where the α, β and γ subunits appear to be the same sizes as the membrane PDE but they are joined by a further δ subunit of 15 kDa.

The analogous PDE found in the cone cells, responsible for colour vision, has a large subunit of 94 kDa joined by three small subunits of 11, 13 and 15 kDa. The overall molecular mass of the complex has been estimated to be approximately 230 kDa.

Some PDEs may in fact be excreted, and are then used to control the levels of extracellular cyclic nucleotides. This would be of importance to organisms such as *Dictyostelium*, which uses the concentrations of extracellular cAMP as a signal for aggregation.

The isolation of a PDE that appears to be specific for cyclic cytidine monophosphate (cCMP) suggests that this third cyclic nucleotide may also have a role in signalling and will no doubt lead to a new interest in this field.

GTPase Superfamily: Functions of Monomeric G Proteins

The role of G proteins, i.e. proteins that bind GDP and GTP and which undergo exchange of the two nucleotides leading to an activated state and hydrolysis of the nucleotide to restore the protein to an inactive state, is far more common than that outlined above for the control of adenylyl cyclase. As well as the plasma membrane-associated trimeric G proteins discussed above, there exists a group of small monomeric G proteins that control such diverse cellular functions as proliferation, protein synthesis, differentiation and regulation of movement of proteins through the cytoplasm. All these G proteins possess intrinsic GTPase activity and the term 'GTPase superfamily' has been coined to encapsulate this growing group of proteins.

As with the trimeric G proteins, the monomeric G proteins are used as a switch mechanism in the cell, with a large amount of signal amplification taking place. The proteins are inactive when bound to GDP, which when exchanged for GTP cause activation and triggering of the signal. The intrinsic GTPase activity converts GTP back to GDP, turning off the signal and allowing the protein to await the next signal (Figure 5.9). It has also been postulated that a third state exists, in which GDP has been removed but GTP has yet to bind. This is a transient state in which either binding of GTP can turn the protein on or binding of GDP can prevent the system turning on. Normally in the cell, the GTP-binding step is favoured. An analogous empty state of the guanine nucleotide-binding site probably also exists in the α-subunit cycle of the trimeric G proteins.

However, certain differences exist between this system and the trimeric G proteins described above. Although these G proteins are monomeric, both the dissociation step, resulting in the loss of GDP, and the GTPase step, converting GTP back to GDP, require the assistance of other proteins. The first, GDP-releasing step is aided by the presence of GNRPs, while GTP hydrolysis is aided by the presence of GTPase-activating proteins (GAPs; see Figure 5.9). It is probable that the α sub-

Figure 5.9
G protein cycle of the monomeric G proteins showing the actions of GTPase-activating protein (GAP) and guanine nucleotide-releasing protein (GNRP)

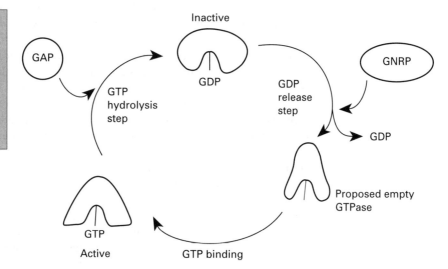

unit of the trimeric G proteins does not need the assistance of a GAP protein for activating the GTPase activity because it contains a region on the polypeptide that has GAP-like activity, i.e. a GAP-like domain.

The classical model of this type of monomeric G protein is the product of the *ras* oncogene, p21^{ras}, or simply Ras. This is a plasma membrane-associated protein of molecular mass 21 kDa. Most of these proteins were first identified as gene products from acute transforming retroviruses. The Ras protein has been crystallised and a structure resolved. Another good model for this system is EF-Tu, elongation factor Tu, involved in protein synthesis. This protein has a molecular mass of approximately 43 kDa and has also been crystallised. Other members of the GTPase family include the proteins Rab, Rho and Rac. At least four Rho polypeptides have been identified along with two Rac polypeptides, as well as other related proteins such as TC10 and CDC42H. Such proteins also work in conjunction with GAP-like proteins, such as Rho-GAP, and GNRP-like proteins, such as Dbl and Rho-GDS.

The role of Ras was originally hinted at by the discovery that transforming genes are in fact alterations of normal genes, activation of which is due to the presence of a point mutation. Their involvement has been suggested in the formation of several forms of cancer as well as in the formation of tumours induced by physical and some chemical reagents. Most of the assays for the presence of Ras depend on the ability of Ras to induce cell proliferation. Bioassays have been based on DNA synthesis, transformation of established cultured cells and the formation of tumours. Other assays have exploited the ability of Ras to induce differentiation of cells. It is interesting to note that full transformation of cells *in vitro* also depends on the presence of another oncogene such as *myc*.

The *ras* gene has been well characterised. In humans and rodents there are three functional *ras* genes present in the genome: H-*ras*, K-*ras* and N-*ras*. H-*ras* and K-*ras* are related to the Harvey (Ha) and Kirsten (Ki) sarcoma viruses of mice. The structure of the gene is unusual in that what appears to be the first 5′ exon is in fact non-coding. The coding region is split into four exons but the K-*ras* gene contains two alternative fourth exons, which can give rise to two separate gene products by alternative splicing. The introns are very different between the various genes, leading to very different lengths of mRNA for each.

The *ras* products have been detected in all tissues explored, including foetal tissue. However, histochemical staining is more intense in cells that are proliferating, as opposed to those which are fully differentiated. Northern blot analysis, which reveals the levels of gene expression, shows that the genes do show some tissue specificity. The control of the levels of transcription is poorly understood for these genes, but a region within the first intron may act as an enhancer, with another small weak enhancer downstream of the gene, while a short alternative exon within another intron also appears to be important, although the exact nature of its role remains obscure. However, the presence of serum or some growth factors can enhance the expression of Ras.

Most Ras proteins are 189 amino acids long, except the K-*ras*B product which is 188 amino acids. The striking feature is that the first 164 amino acids are homologous across all *ras* products, with the C-terminal region being referred to as the heterogeneous region due to its divergence. However, even here sequence patterns are conserved between species, suggesting that certain C-terminal sequences confer functional specificity to the proteins. The first 164 amino acids probably contain the GTPase activity of the protein. In all Ras proteins a cysteine residue is conserved at a position four residues before the C-terminus. The protein is synthesised as a pro-p21^{ras} product and this C-terminal region also undergoes a great deal of post-translational modification, including farnesylation, cleavage of the terminal amino acids and methylation (see Chapter 4). These modifications are involved in the increase in the hydrophobicity of the protein that encourages its association with the plasma membrane. Mutant proteins without this conserved cysteine are cytosolic and have no transforming activity.

The general structure of the p21^{ras} protein, Ras, shows that it has a hydrophobic core consisting of six β-sheet strands connected by hydrophilic loops and α-helices. Five regions of the protein, which have been designated G1–G5, lie on one side of the protein and are associated with the hydrophilic loops. G1 is probably involved in the binding of the first two phosphate groups, α and β, of GDP and GTP. The catalytic step of the cycle requires the presence of Mg^{2+} and is probably associated with regions G2 and G3. The G2 and G3 regions are probably the site of conformational changes induced by the binding of GTP and so activation of the protein. They have also been implicated in the association with GAP proteins. The other association needed for the functioning of the GTPase activity is interaction with the subsequent protein along the pathway, i.e. the effector. Here again, the GTPase region G2 has been implicated.

Ras oncogenes, which have the ability to promote tumour formation, can be used to determine the active residues in the protein. Commonly, substitutions of residues at positions 12 (glycine) and 61 (glutamine) reduces the intrinsic GTPase activity of the protein. It is now believed that the catalytic cycle involves a glutamine residue on the protein, which activates a water molecule that then attacks the γ-phosphoryl group of GTP in a nucleophilic fashion. However different GTPases almost certainly vary in the precise mechanism of GTP hydrolysis.

Some viral *ras* genes encode a protein with threonine at position 59. Interestingly, this protein product undergoes autophosphorylation of this threonine, but the role of this phosphoryl addition has not been determined. Another fascinating mutant form of Ras has no ability to bind to guanine nucleotides and therefore its conformation must be permanently locked into an active form.

Let us now turn to the proteins that interact with the monomeric G proteins. Several proteins have been found to contain GAP activity. One of these is a protein with an approximate molecular mass of 120 kDa (p120GAP). It consists of 1047 amino acids, and appears to be expressed everywhere. The activity of the protein, (i.e. its ability to increase the GTPase activity of the Ras protein up to fivefold) is due to a catalytic domain in the C-terminal end of the protein. The polypeptide also contains several protein-binding domains: two SH2 domains, an SH3 domain and a PH domain, which must be involved in its interaction with other polypeptides. Two such polypeptides, which are themselves phosphorylated, include one of 190 kDa (p190) and one of 62 kDa (p62). p120GAP complexed with p190 has a reduced ability to stimulate GTPase activity of Ras while p62 has been implicated in the control of mRNA processing. p120GAP is also phosphorylated on tyrosine residues by tyrosine kinase receptors, which therefore probably have a measure of control on its activity, and it is inhibited by some lipids, including arachidonic acid.

A second GAP protein is the family known as GAP1. Mammals contain at least two homologues, GAP1^m and GAP1^{IP4BP}. These proteins, of approximately 850 amino acids, contain a PH domain towards the C-terminal end of the polypeptide, a GAP-related domain known as the GRD domain in the middle and two domains homologous to the C$_2$ regulatory domains of PKC towards the N-terminus. Interestingly, these proteins have been shown to bind to inositol 1,3,4,5-tetrakisphosphate (InsP$_4$), one of the metabolites derived from inositol 1,4,5-trisphosphate (InsP$_3$), which suggests exciting possibilities in its control. Like the other GAP proteins these proteins also show inhibition by phospholipids. Expression of this protein is not as widespread as for p120GAP, with the highest expression seen in the placenta, brain and kidneys.

GAPs are not the only proteins found to stimulate the GTPase activity of Ras. A protein of approximately 250 kDa known as NF1, otherwise called neurofibromin, is responsible for Recklinghausen's neurofibromatosis (NF1 disease). Of its 2818 amino acids, a region of 350 amino acids in the centre section of the polypeptide is analogous to GAP. The protein appears to be expressed primarily in the nervous system, particularly in neurones and Schwann cells, and exists in at least three isoforms, arising from alternative splicing. The role of neurofibromin almost certainly involves its interaction with other proteins in a complex and again its activity is inhibited by certain lipids.

Other members of the Ras superfamily are not necessarily acted upon by GAP and have their own proteins that are analogous to GAP. However, the GAP1 family, as discussed above, does show GAP activity against the Rap G protein, although this activity is not affected by lipids or interestingly by the presence of InsP$_4$.

The other G protein regulatory proteins are the GNRPs, which include the protein known as Sos. This protein is often found associated with an adaptor protein, such as GRB2 in mammals or Drk in *Drosophila*, the interaction being possibly through the adaptor's SH3 domains. On binding of a ligand to its respective receptor protein kinase, the receptor usually can autophosphorylate and the formation of the new phosphotyrosine groups allows interaction with an adaptor protein through its SH2 domains. This ensures that the adaptor protein and its associated GNRP, for example Sos, is now in close association with the membrane where the latter leads to nucleotide exchange and activation of the G protein. Such a scheme is illustrated in Figures 4.9 and 9.4.

Other GNRPs include the gene product of *cdc25* in *S. cerevisiae*, a 1545-amino acid polypeptide. Mutants of *S. cerevisiae* lacking the CDC25 protein, *cdc25*$^-$ mutants, have been useful in the identification of similar proteins. An analogous gene in mammals has been discovered, *CDC25*^{Mm}, which codes for a protein of 140 kDa found in brain tissue. However it is another complex gene, which can give rise to at least four distinct proteins of very disparate molecular masses. Further, a protein of 35 kDa has been found to stimulate guanine nucleotide exchange, while another of cytosolic origin, smgp21GDS, has been found to be active only if Ras has undergone its post-translational modifications. To further complicate the issue, inhibitor proteins of guanine nucleotide exchange have been reported that act on several members of the Ras superfamily, particularly those of the Rab, Rho and Rac families, for example Rho-GDI.

Having discovered the presence of these monomeric G proteins, their role in cell signalling can only be fully understood if it is found what causes their activation and how they fit into transduction pathways. The activity of Ras can be increased in one of two ways. Either the GTPase activity is decreased, for example through a decrease of GAP activity, or GTP/GDP exchange is increased, through the increase of GNRP activity. It has been shown that the transformation abilities of several oncogenes encoding protein tyrosine kinases, for example *src* and *fms*, require the activation of Ras. Good examples of Ras involvement in transduction pathways are insulin-induced signalling (discussed in Chapter 9) and the EGF-induced pathway (see Figure 4.9). Here Ras itself is found to activate the serine/threonine kinase, Raf, an oncogene product, which leads to the activation of MAPK cascades. However, a couple of reports have shown that Raf activation can also be independent of Ras if the Raf protein is associated with the plasma membrane.

Ras can also activate other serine/threonine kinases and has also been reported to activate MAPKKs as well as MAPKs. Activation of ribosomal protein S6 kinase as well as PKC has also been seen.

As with the trimeric G proteins, Ras has been shown to activate the enzyme adenylyl cyclase in *S. cerevisiae*. However the effector role of Ras may be mediated through GAP proteins and neurofibromin. Binding of Ras to GAPs is thought to alter the conformation of the GAP proteins, leading to the possible exposure of protein-interacting domains, such as SH2 and SH3 domains. Once exposed, these domains would enable GAP proteins to interact with other proteins and so propagate a signal within the cell, without the direct involvement of Ras.

Other Ras-related Proteins

A subgroup of the Ras family of proteins is the ARFs (ADP-ribosylation factors). These were first identified as factors needed for the efficient modification of the trimeric G proteins $G_{s\alpha}$ by cholera toxin. These proteins have been shown to be involved in the vesicle-mediated protein traffic of cells as well as in the control of some lipid signalling by regulation of the activity of phospholipase D (PLD), which yields phosphatidic acid (PA) and choline from membrane lipids.

The elongation factor involved in protein synthesis, EF-Tu is also very similar to Ras in its mode of action. However, the GNRP involved in GDP/GTP exchange is another elongation factor, EF-Ts. This protein contains constitutive activity, unlike the trimeric exchange which involves the activation of a receptor.

Summary

Cyclic nucleotides are central signals in many transduction pathway (Figure 5.10). cAMP is produced from ATP by the enzyme adenylyl cyclase. In mammals this is seen as a single polypeptide integral to the plasma membrane, but many other

Figure 5.10

A simplified diagram showing the role of cAMP and G proteins in the transduction of cellular signals. DAG, diacyglycerol; GAP, GTPase-activating protein; GNRP, guanine nucleotide-releasing protein; InsP$_3$, inositol trisphosphate.

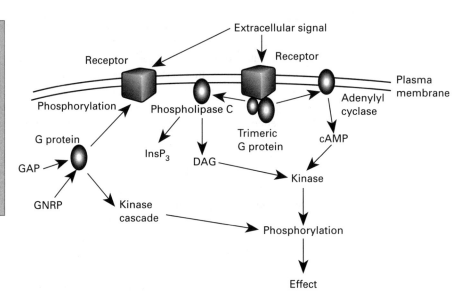

isoforms are seen in nature. The main control of the enzyme's activity is through its interaction with the trimeric class of G proteins. These are constructed of an α subunit, which can either stimulate ($G_{s\alpha}$) or inhibit ($G_{i\alpha}$) adenylyl cyclase, along with a β/γ subunit complex. Although once thought to be inert, the β/γ complex itself is postulated to fulfil a signalling role.

The trimeric G proteins are a wide family and include members that interact with PDEs (G_t) and regulate PLC (G_q).

cGMP is produced by an enzyme analogous to that producing cAMP, but here two forms are seen, a soluble type and a membrane-bound type.

The cyclic nucleotide signal is ended by hydrolysis of the cyclic forms to the monophosphate forms by PDEs. Seven classes of these enzymes have been reported, including ones which are controlled by Ca^{2+}/calmodulin, stimulated by cGMP and inhibited by cGMP.

Another important group of signalling molecules is the monomeric G proteins, which include the oncogene products Ras and Rac as well as the protein synthesis elongation factor EF-Tu. Like the trimeric forms, these G proteins are active when bound to GTP and their intrinsic GTPase activity hydrolyses GTP to GDP accompanied by a conformational change in the protein and its return to the inactive state. This activity of the monomeric proteins is enhanced by the protein GAP, while GDP release is enhanced by the GNRPs. These G proteins are instrumental in the transduction of the signal from receptors on the membrane to kinase cascades such as MAPKs. Another group of proteins, the adaptor proteins, is also likely to be involved here.

Further reading

Adenylyl Cyclase

Cooper, D. M. F., Mons, N. and Karpen, J. W. (1995) Adenylyl cyclases and the interaction between calcium and cAMP signalling. *Nature*, **374**, 421–4.

Krupinski, J., Coussen, F., Bakalyar, H. A. *et al.* (1989) Adenylyl cyclase amino acid sequence: possible channel-like or transporter-like structure. *Science*, **244**, 1558–64.

Tang, W.-J. and Gilman, A. G. (1991) Type specific regulation of adenylyl cyclase by G protein beta/gamma subunits. *Science*, **254**, 1500–3.

Tang, W.-J. and Gilman, A. G. (1992) Adenylyl cyclases. *Cell*, **70**, 869–72.

Tang, W.-J., Krupinski, J. and Gilman, A. G. (1991) Expression and characterisation of calmodulin activated type I adenylyl cyclase. *Journal of Biological Chemistry*, **266**, 8595–603.

Adenylyl Cyclase Control and the Role of G Proteins

Bauer, P. H., Muller, S., Puzicha, M. *et al.* (1992) Phosducin is a protein kinase A regulated G protein regulator. *Nature*, **358**, 73–6.

Birnbaumer, L. (1992) Receptor-to-effector signalling through G proteins: roles of βγ dimers as well as a subunits. *Cell*, **71**, 1069–72.

Camps, M., Hou, C. F., Sidiropoulos, D., Stock, J. B., Jakobs, K. H. and Gierschik, P. (1992) Stimulation of phospholipase C by guanine nucleotide binding protein beta/gamma subunits. *European Journal of Biochemistry*, **206**, 821–31.

Hepler, J. R. and Gilman, A. G. (1992) G proteins. *Trends in Biochemical Sciences*, **17**, 383–7.

Igarashi, M., Strittmatter, S. M., Vartanian, T. and Fishman, M. C. (1993) Mediation by G proteins of signals that cause collapse of growth cones. *Science*, **259**, 77–9.

Iñiguz-Lluhi, J., Kleuss, C. and Gilman A. G. (1993) The importance of G-protein β/γ subunits. *Trends in Cell Biology*, **3**, 230–5.

Landis, C. A., Masters, S. B., Spada, A., Pace, A. M., Bourne, H. R. and Vallar, L. (1989) GTPase inhibiting mutations activate the α-chain of G_s and stimulate adenylyl cyclase in human pituitary tumors. *Nature*, **340**, 692–6.

Leberer, E., Dignard, D., Hougan, L., Thomas, D. Y. and Whiteway, M. (1992) Dominant negative mutants of a yeast G protein beta subunit identify two functional regions involved in pheromone signalling. *EMBO Journal*, **11**, 4805–13.

Lee, R. H., Ting, T .D., Lieberman B. S., Tobias, D. E., Lolley, R. N. and Ho, Y. K. (1992) Regulation of retinal cGMP cascade by phosducin in bovine rod photoreceptor cells: interaction of phosducin and transducin. *Journal of Biological Chemistry*, **267**, 25104–12.

Linder M. E. and Gilman, A. G., (1992) G proteins. *Scientific American*, **267**, 36–43.

Linder, M. E., Pang, I. H., Duronio, R. J., Gordon, J. I., Sternweis, P. C. and Gilman, A. G. (1991) Lipid modifications of G protein subunits: myristoylation of G alpha increases its affinity for the beta-gamma. *Journal of Biological Chemistry*, **266**, 4654–9.

Logotheetis, D. E., Kurachi, Y., Galper, J., Neer, E. J. and Clapham, D. E. (1987) The beta subunit and gamma subunit of GTP binding proteins activate the muscarinic K^+ channel in heart. *Nature*, **325**, 321–6.

Masters, S. B., Sullivan, K. A., Miller, R. T. *et al.* (1988) Carboxyl terminal domain of G_s alpha specifies coupling of receptors to stimulation of adenylyl cyclase. *Science*, **241**, 448–51.

Mattera, R., Graziano, M. P., Yatani, A. *et al.* (1989) Splice variants of the alpha subunit of the G protein G_s activate both adenylyl cyclase and calcium channels. *Science*, **243**, 804– 7.

Moss, J. and Vaughan, M. (eds.) (1990) *ADP-ribosylating Toxins and G Proteins: Insights in Signal Transduction*. American Society for Microbiology, Washington.

Pang, I. H. and Sternweis, P. C. (1990) Purification of unique α subunits of GTP binding regulatory proteins (G proteins) by affinity chromatography with immobilised β/γ subunits. *Journal of Biological Chemistry*, **265**, 18707–12.

Pitcher, J. A., Inglese, J., Higgins, J. B. *et al.* (1992) Role of beta-gamma subunits of G proteins in targeting the β-adrenergic receptor kinase to membrane bound receptors. *Science*, **257**, 1264–7.

Pitt, G. S., Gunderson, R. E., Lilly, P. J., Pupillo, M. B. and Vaughan, R. A. (1990) G protein-linked signal transduction in aggregating *Dictyostelium*. In *G Proteins and Signal Transduction*, Nathanson, N. M. and Harden, T. K. (eds). The Rockefeller University Press, New York.

Schubert, B., VanDongen, A. M. J., Kirsch, G. E. and Brown, A. M. (1989) β-Adrenergic inhibition of cardiac sodium channels by dual G protein pathways. *Science*, **245**, 516–19.

Simon, M. I., Strathmann, M. P. and Gautam, N. (1991) Diversity of G proteins in signal transduction. *Science*, **252**, 802–8.

Guanylyl Cyclase

Ashman, D. F., Lipton, R., Melicow, M. M. and Price, T. D. (1963) Isolation of adenosine 3′,5′-monophosphate and guanosine 3′,5′-monophosphate from rat urine. *Biochemical and Biophysical Research Communications*, **11**, 330–4.

Chinkers, M., Singh, S. and Garbers, D. L. (1991) Adenine nucleotides are required for activation of rat atrial natriuretic peptide receptor guanylyl cyclase expressed in a baculovirus system. *Journal of Biological Chemistry*, **266**, 4088–93.

Corbin, J. D., Thomas, M. K., Wolfe, L., Shabb, J. B., Woodford, T. A. and Francis, S. H. (1990) New insights into cGMP action. In *Advances in Second Messenger and Phosphoprotein Research: The Biology and Medicine of Signal Transduction*, Nishizuka, Y., Endo M. and Tanaka, C. (eds), vol. **24**, pp. 411–18. Raven Press, New York.

Dizhoor, A. M., Ray, S., Kumar, S. *et al.* (1991) Recoverin: a calcium sensitive activator of retinal rod guanylate cyclase. *Science*, **251**, 915–18.

Dizhoor, A. M., Lowe, D. G., Olshevskaya, E. V., Laura, R. P. and Hurley, J. B. (1994) The human photoreceptor membrane guanylyl cyclase, RETGC, is present in outer segments and is regulated by calcium and a soluble activator. *Neuron*, **12**, 1345–52.

Garbers, D. L. (1989) Guanylyl cyclase: a cell surface receptor. *Journal of Biological Chemistry*, **264**, 9103–6.

Garbers, D. L. and Lowe, D. G. (1994) Guanylyl cyclase receptors. *Journal of Biological Chemistry*, **269**, 30741–44.

Gerzer, R., Bohme, E., Hofmann, F. and Schultz, G. (1981) Soluble guanylate cyclase purified from bovine lung contains heme and copper. *FEBS Letters*, **132**, 71–4.

Goldberg, N. D., Ames, A., Gander, J. E., and Walseth, T. F. (1983) Magnitude of increase in retinal cGMP metabolic flux determined by ^{18}O incorporation into nucleotide α-phosphoryls corresponds with intensity of photic stimulation. *Journal of Biological Chemistry*, **258**, 9213–19.

Goraczniak, R. M., Duda, T., Sitaramayya, A. and Sharma, R. K. (1994) Structural and functional characterisation of the rod outer segment membrane guanylyl cyclase. *Biochemical Journal*, **302**, 455–61.

Gorczyca, W. A., Gray-Keller, M. P., Detwiler, P. B. and Palczewski, K. (1994) Purification and physiological evaluation of a guanylate cyclase activating protein from retinal rods. *Proceedings of the National Academy of Science USA*, **91**, 4014–18.

Horio, Y. and Murad, F. (1991) Solubilisation of guanylyl cyclase from bovine rod outer segments and effects of lowering Ca^{2+} and nitro compounds. *Journal of Biological Chemistry*, **266**, 3411–15.

Idriss, S. D., Pilz, R. B., Sharma, V. S and Boss, G. R. (1992) Studies on cytosolic guanylate cyclase from human placenta. *Biochemical and Biophysical Research Communications*, **183**, 312–20.

Nakane, M., Arai, K., Saheki, S., Kuno, T., Buechler, W. and Murad, F. (1990) Molecular cloning and expression of cDNAs coding for soluble guanylate cyclase from rat lung. *Journal of Biological Chemistry*, **265**, 16841–5.

Severina, I. S. (1992) Soluble guanylate cyclase of platelets: function and regulation in normal and pathological states. *Advances in Enzyme Regulation*, **32**, 35–56.

Shyjan, A. W., de Sauvage, F. J., Gillet, N. A., Goeddel, D. V. and Lowe, D. G. (1992) Molecular cloning of a retinal specific membrane guanylyl cyclase. *Neuron*, **9**, 727–37.

Stone, J. R. and Marletta, M. A. (1993) Bovine lung contains multiple isoforms of soluble guanylate cyclase. *FASEB Journal*, **7**, A1152.

Yuen, P. S. T. and Garbers, D. L. (1992) Guanylyl cyclase-linked receptors. *Annual Review of Neuroscience*, **15**, 193–225.

Phosphodiesterases

Beavo, J. A. (1988) Multiple isozymes of cyclic nucleotide phosphodiesterase. *Advances in Second Messenger and Phosphoprotein Research*, **22**, 1–38.

115

Deterre, P., Bigay, J., Forquet, F., Robert, M and Chabre, M. (1988) cGMP phosphodiesterase of retinol rods is regulated by 2 inhibitory subunits. *Proceedings of the National Academy of Sciences USA*, **85**, 2424–8.

Gillespie, P. G. and Beavo, J. A. (1988) Characterisation of a bovine cone photoreceptor phosphodiesterase purified by cyclic GMP-sepharose chromatography. *Journal of Biological Chemistry*, **263**, 8133–41.

Harrison, S. A., Reifsnyder, D. H., Gallis, B., Cadd, G. G. and Beavo, J. A. (1986) Isolation and characterisation of bovine cardiac muscle cGMP-inhibited phosphodiesterase: a receptor for new cardiotonic drugs. *Molecular Pharmacology*, **29**, 506–14.

Kasuya, J., Goko, H. and Fujita-Yamaguchi, Y. (1995) Multiple transcripts for the human cardiac form of the cGMP-inhibited cAMP phosphodiesterase. *Journal of Biological Chemistry*, **270**, 14305–12.

Manganiello, V. C., Smith, C. J., Degerman, E. and Belfrage, P. (1990) Cyclic GMP-inhibited cyclic nucleotide phosphodiesterases. *Cyclic Nucleotide Phosphodiesterases: Structure, Regulation and Drug Design*, Beavo, J. and Houslay, M. D. (eds), pp. 87–116. John Wiley and Sons, Chichester.

Martins, T. J., Mumby, M. C. and Beavo, J. A. (1982) Purification and characterisation of a cyclic GMP-stimulated cyclic nucleotide phosphodiesterase from bovine tissue. *Journal of Biological Chemistry*, **257**, 1973–9.

Milatovich, A., Bolger, G., Michaeli, T. and Francke, U. (1994) Chromosome localizations of genes for five cAMP-specific phosphodiesterases in man and mouse. *Somatic Cell and Molecular Genetics*, **20**, 75–86.

Monaco, L., Vicini, F. and Conti, M. (1994) Structure of two rat genes coding for closely related rolipram sensitive cAMP phosphodiesterases: multiple messenger RNA variants originate from alternative splicing and multiple start sites. *Journal of Biological Chemistry*, **269**, 347–57.

Newton, R. P. and Salih, S. G. (1986) Cyclic CMP phosphodiesterase: isolation, specificity and kinetic properties. *International Journal of Biochemistry*, **18**, 743–752.

GTPase Superfamily

Barbacid, M. (1987) Ras genes. *Annual Review of Biochemistry*, **56**, 779–827.

Bourne, H. R., Sanders, D. A. and McCormick, F. (1991) The GTPase superfamily: conserved structure and molecular mechanism. *Nature*, **349**, 117–27.

Cen, H., Papageorge, A., Zippel, R., Lowy, D. R. and Zhang, K. (1992) Isolation of multiple mouse cDNAs with coding homology to *Saccharomyces cerevisiae* CDC25: identification of a region related to BRB, VAL, DBL and CDC24. *EMBO Journal*, **11**, 4007–15.

Cohen, J. B. and Levinson, A. D. (1988) A point mutation in the last intron responsible for increased expression of transformed activity of the c-Ha Ras oncogene. *Nature*, **334**, 119–24.

Cullen, P. J., Hauan, J. J., Truong, O. *et al.* (1995) Identification of a specific Ins(1,3,4,5)P$_4$ binding protein as a member of the GAP1 family. *Nature*, **376**, 527–30.

Frech, M., John, J., Pizon, V. *et al.* (1990) Inhibition of GTPase activating protein stimulation of Ras-P21 GTPase by the *Krev*-1 gene product. *Science*, **249**, 169–71.

Fukuda, M. and Mikoshiba, K. (1996) Structure–function relationships of the Mouse Gap1^m: determination of the inositol 1,3,4,5-tetrakisphosphate binding domain. *Journal of Biological Chemistry*, **271**, 18838–42.

Hancock, J. F., Magee, A. I., Childs, J. E. and Marshall, C. J. (1989) All Ras proteins are polyprenylated but only some are palmitoylated. *Cell*, **57**, 1167–77.

Hancock, J. F., Cadwallader, K. and Marshall, C. J. (1991) Methylation and proteolysis are essential for efficient membrane binding of phenylated p21K-Ras(B). *EMBO Journal*, **10**, 641–6.

Hata, Y., Kikuchi, A., Sasaki, T., Schaber, M. D., Gibbs, J. B. and Takai, Y. (1990) Inhibition of the Ras-P21 GTPase activating protein stimulated GTPase activity of c-Ha-Ras-P21 by Smg-P21 having the same putative effector domain as Ras-P21s. *Journal of Biological Chemistry*, **265**, 7104–7.

Hiraoka, K., Kaibuchi, K., Ando, S. *et al.* (1992) Both stimulatory and inhibitory GDP/GTP exchange proteins, Smg-GDS and Rho-GDI, are active on multiple small GTP binding proteins. *Biochemical and Biophysical Research Communications*, **182**, 921–30.

Lowy, D. R. and Willumsen, B. M. (1993) Function and regulation of Ras. *Annual Review of Biochemistry*, **62**, 851–91.

Marchuk, D. A., Saulino, A. M., Tavakkol, R. *et al.* (1991) cDNA cloning of the type-1 neurofibromatosis gene: complete sequence of the NF1 gene product. *Genomics*, **11**, 931–40.

Medema, R. H., Delaat, W. L., Martin, G. A., McCormick, F. and Bos, J. L. (1992) GTPase activating protein SH2–SH3 domains induce gene expression in a Ras-dependent fashion. *Molecular and Cellular Biology*, **12**, 3425–30.

Milburn, M. V., Tong, L., Devos, A. M. *et al.* (1990) Molecular switch for signal transduction: structural differences between active and inactive forms of the proto-oncogene Ras proteins. *Science*, **247**, 939–45.

Moran, M. F., Polakis, P., McCormick, F., Pawson, T. and Ellis, C. (1991) Protein tyrosine kinases regulate the phosphorylation, protein interactions, subcellular distribution and activity of p21 Ras GTPase activating protein. *Molecular and Cellular Biology*, **11**, 1804–12.

Morris, J. D. H., Price, B., Lloyd, A. C., Self, A. J., Marshall, C. H. J. and Hall, A. (1989) Scrape-loading of Swiss 3T3 cells with Ras protein rapidly activates protein kinase C in the absence of phosphoinositide hydrolysis. *Oncogene*, **4**, 27–31.

Nishi, T., Lee, P. S. Y., Oka, K. *et al.* (1991) Differential expression of two types of the neurofibromatosis type-1 (NF1) gene transcripts related to neuronal differentiation. *Oncogene*, **6**, 1555–59.

Pai, E. F., Kabsch, W., Krengel, U., Holmes, K. C., John, J. and Wittinghofer, A. (1989) Structure of the guanine nucleotide binding domain of the Ha–Ras oncogene product P21 in the triphosphate conformation. *Nature*, **341**, 209–14.

Pai, E. F., Krengel, U., Petsko, G. A., Goody, R. S., Kabsch, W. and Wittinghofer, A. (1990) Refined crystal structure of the triphosphate conformation of H-Ras

P21 at 1.35Å resolution: implications for the mechanism of GTP hydrolysis. *EMBO Journal*, **9**, 2351–9.

Ruley, H. E. (1990) Transforming collaborations between Ras and nuclear oncogenes. *Cancer Cells: A Monthly Review*, **2**, 258–68.

Sasaki, T., Kaibuchi, K., Kabcenell, A. K., Novick, P. J. and Takai, Y. (1991) A mammalian inhibiting GDP/GTP exchange protein (GDP dissociation inhibitor) for Smg P25A is active on yeast Sec4 protein. *Molecular and Cellular Biology*, **11**, 2909-12.

Schlessinger, J. (1993) How receptor tyrosine kinases activate Ras. *Trends in Biochemical Sciences*, **18**, 273–5.

Schlichting, I., Almo, S. C., Rapp, G. *et al.* (1990) Time resolved X-ray crystallographic study of the conformational change in Ha-Ras P21 protein on GTP hydrolysis. *Nature*, **345**, 309–15.

Shou, C., Farnsworth, C. L., Neel., B. G. and Feig, L. A. (1992) Molecular cloning of cDNA encoding a guanine nucleotide releasing factor for Ras P21. *Nature*, **358**, 351–4.

Sigal, I. S., Gibbs, J. B., D'Alonzo, J. S. and Scolnick, E. M. (1986) Identification of effector residues and a neutralising epitope of Ha-Ras encoded P21. *Proceedings of the National Academy of Sciences USA*, **83**, 4725–9.

Trahey, M. and McCormick, F. (1987) A cytoplasmic protein stimulates normal N-Ras P21 GTPase but does not affect oncogenic mutants. *Science*, **238**, 542–5.

Vojtek, A. B., Hollenberg, S. M. and Cooper, J. A. (1993) Mammalian Ras interacts directly with the serine threonine kinase Raf. *Cell* , **74**, 205–14.

Other Ras-related Proteins

Brown, H. A., Gutowki, S., Moomaw, C. R., Slaughter, C. and Sternweis, P. C. (1993) ADP-ribosylation factor, a small GTP-dependent regulatory protein, stimulates phospholipase D activity. *Cell*, **75**, 1137–44.

Romero, G., Chau, V. and Biltonen, R.L. (1985) Kinetics and thermodynamics of the interaction of elongation factor Tu with elongation factor Ts, guanine nucleotides and aminoacyl transfer RNA. *Journal of Biological Chemistry*, **260**, 6167–74.

6 Inositol Phosphate Metabolism and the Roles of Other Membrane Lipids

Introduction

One of the key events in signal transduction in cells takes place on the plasma membrane and involves the breakdown of some of the lipids that comprise the membrane. Between 2 and 8% of the lipids of eukaryotic membranes are inositol-containing lipids. The three main forms of these lipid structures are phosphatidylinositol (PtdIns), phosphatidylinositol 4-phosphate (PtdIns 4-P) and $PtdInsP_2$, with the non-phosphorylated form, PtdIns, accounting for more than 80% of the total inositol lipid content. Signalling can occur because key proteins on the membrane are responsible for the cleavage of these lipid molecules into smaller, diffusible molecules that convey the signal to other parts of the cell. Primarily, $PtdInsP_2$ is broken down by the enzyme PLC to give the primary products, $InsP_3$ and DAG. This pathway, like others, greatly amplifies the signal, as one activated PLC molecule produces many $InsP_3$ and DAG molecules. Further, the system leads to great divergence of the signal. $InsP_3$ is responsible for the release of Ca^{2+} ions from intracellular stores, which leads to the activation of the Ca^{2+} of the signalling pathways, including the activation of calmodulin and its associated effects (discussed in the next chapter), while DAG leads to the activation of PKC and the associated phosphorylation of a host of proteins along with modulation of their activity (discussed in Chapter 4). A typical scheme is depicted in Figure 6.1. Therefore, inositol phosphate metabolism can be seen as a keystone pathway linking events at the membrane with other transduction pathways deep within a cell. To begin with, it was thought that this was the end of the story, but a plethora of reports have indicated that it is just not that simple. More than one type of membrane-associated lipid can be cleaved, while $InsP_3$ can lead to a host of secondary products and, gradually, many of these have been assigned signalling roles within the cell.

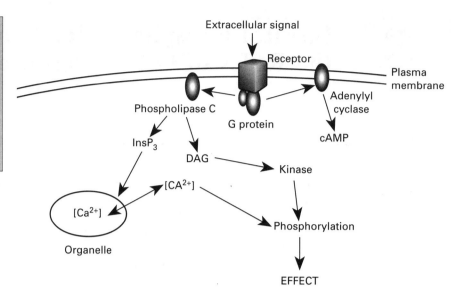

Figure 6.1
Control and role of phospholipase C in inositol phosphate metobolism and its influence on kinase activity and intracellular Ca^{2+} release. DAG, diacylglycerol; $InsP_3$, inositol trisphosphate.

At this point it is probably worth mentioning nomenclature. Here inositol 1,4,5-trisphosphate is referred to as $InsP_3$, but many reports now find that a more complete abbreviation is needed to clarify which trisphosphate is being referred to, and a similar story goes for other phosphorylated derivatives. For example, phosphatidylinositol bisphosphate could be phosphorylated at positions 3 and 4 or at positions 4 and 5. However, to try to simplify the situation I have used a minimalist approach and the abbreviation refers to the most important isomers recorded, i.e. $InsP_3$ for inositol 1,4,5-trisphosphate, $PtdInsP_2$ for phosphatidylinositol 4,5-bisphosphate, while other derivatives that have a less well-defined role within the cell are given a more complete name with numbers as well for clarity. It should also be noted that some researchers and journals use another type of nomenclature, the 'Chilton' forms, i.e. $InsP_3$ is known as IP_3 and $PtdInsP_2$ as PIP_2. Conventions have been published giving guidelines to which nomenclature to use, but many journals including the *Biochemical Journal* still accept publication of both forms, for example $InsP_3$ or IP_3.

Events at the Membrane

The main lipid cleaved in the membrane is $PtdInsP_2$, which leads to the production of $InsP_3$ and DAG. The hydrophobic tails of the lipid comprise part of the inner leaflet of the membrane bilayer, while the phosphates and inositol group are sticking out into the cytoplasm of the cell (Figure 6.2). $PtdInsP_2$ is itself produced by the sequential phosphorylation of PtdIns. PtdIns is first phosphorylated to PtdIns 4-P by PtdIns 4-

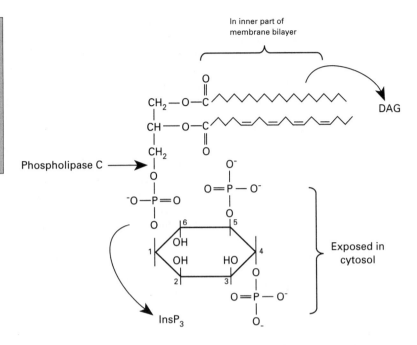

Figure 6.2
Molecular structure of phosphatidylinositol 4,5-bisphosphate and its breakdown by phospholipase C to form inositol trisphosphate (InsP$_3$) and diacylglycerol (DAG).

kinase and then further modified by the addition of a phosphoryl group to the 5 position of the inositol ring, catalysed by PtdIns 4-P 5-kinase, to produce PtdInsP$_2$. The former enzyme, PtdIns 4-kinase, has been purified from several sources, including liver and brain, and in most tissues is associated with membranes. PtdIns 4-P 5-kinase, on the other hand, has been found in both the particulate and soluble fractions of cells. In erythrocytes, two immunologically distinct forms of the enzyme have been found, a 53 kDa form that is both soluble and membrane associated and a second form only found in membranes.

However, like all of the inositol story, it does not end there. Several PtdIns derivatives phosphorylated on the 3 position of the inositol ring were postulated to exist (Figure 6.3). These were first identified in fibroblasts by Whitman and colleagues in

Figure 6.3
Phosphorylated forms of phosphatidylinositol (PtdIns) that may be formed and their interconversion. DAG, diacylglycerol; InsP$_3$, inositol trisphosphate.

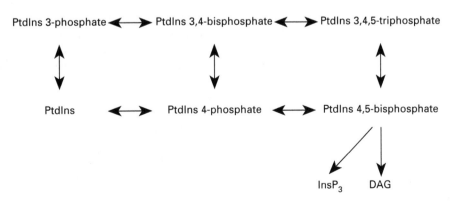

1988, and it was suggested that they were involved in the control of cell proliferation. However, their presence in cells that are differentiated and do not undergo proliferation suggests that other functions also need to be assigned to these lipids. They are interconvertible by simple phosphorylation and dephosphorylation steps and this has led to the hypothesis that some of these lipids are themselves signalling molecules. For example, it is thought that phosphatidylinositol 3,4,5-trisphosphate, derived mainly from the phosphorylation of $PtdInsP_2$, may be involved in the activation of the small G protein Rac. The kinase involved in the production of these 3-phospho derivatives of the inositol lipids, PtdIns 3-kinase, is activated by tyrosine kinase activity, such as RTKs, and recent evidence has also suggested that its activation can be mediated by the monomeric G protein Ras. Some reports also show activation of PtdIns 3-kinase by the multiphosphorylated protein IRS-1. Research into the role of PtdIns 3-kinase has been greatly facilitated by the finding that it can be reasonably specifically inhibited by the fungal metabolite, wortmannin.

The lipid $PtdInsP_2$, as well as being the main source of DAG and $InsP_3$, is involved in interactions with proteins associated with the cytoskeleton. Interactions with gelsolin and profilin have been characterised and found to be important in the control of cytoskeletal formation.

Breakdown of the Inositol Phosphate Lipids: Phospholipase C

The hydrolysis of $PtdInsP_2$ in the lipid bilayer of the membrane is catalysed by the enzyme PLC. This enzyme also hydrolyses the other inositol lipids PtdIns and PtdIns 4-P. This reaction releases the inositol phosphates such as $InsP_3$ and DAG (Figure 6.2).

Both cloning and protein work has revealed the existence of many isoforms of PLC. In mammalian tissues for example, nine isoforms have been identified that can be classified into four main groups, α, β, γ and δ. Little sequence homology is seen between the β, γ and δ isoforms, although two homology domains have been identified. The smaller of these domains, or X domain, is approximately 170 amino acids while the other, the Y domain, is larger at approximately 260 amino acids. As the substrates for all these enzymes are within the membrane bilayer it is assumed that they must be membrane associated, although PLCα is the only form that appears to contain a membrane-spanning region. However, PLCγ has an N-terminal region that contains a Ca^{2+}-dependent phospholipid-binding sequence.

Activation of PLC is probably brought about in different ways depending on the isoform. One of the mechanisms for turning on PLC is through the interaction with components of the trimeric G proteins. It has been found for example that PLCβ1 is activated by the α subunit of the trimeric G protein G_q, i.e. $G_{q\alpha}$ (Figure 6.4). Two sites, referred to as P and G, at the C-terminal end of PLCβ1 are important for this interaction. However, this G-protein subunit has no stimulatory effect on any of the other forms of PLC, although the β/γ subunit complex of the trimeric G protein family does activate the β2 and β3 isoforms of PLC in HL-60 promyeloid cells.

It has also been reported that Ca^{2+} may also be involved in PLC control. PLCδ contains one EF hand motif, which confers Ca^{2+}-binding potential to this protein, and therefore PLCδ may represent a Ca^{2+}-controlled form of the PLC family.

Phosphorylation is also crucial in the activation of other isoforms. PLCγ is phosphorylated on some tyrosine residues, usually those at positions 771, 783 and 1254. This may be catalysed by a tyrosine kinase linked receptor, for example the EGF receptor (Figure 6.4). Experiments with PTPs suggest that this is an important step

Figure 6.4
Activation of phospholipase C may be by different mechanisms depending on the isoform. (a) Activation may be through a receptor linked to a trimeric G protein, where the α subunit activates phospholipase, as with phospholipase Cβ1. (b) Activation may be through a tyrosine kinase-linked receptor, leading to phosphorylation of phospholipase C on tyrosine residues, as with phopholipase Cγ. DAG, diacylglycerol; InsP$_3$, inositol trisphosphate.

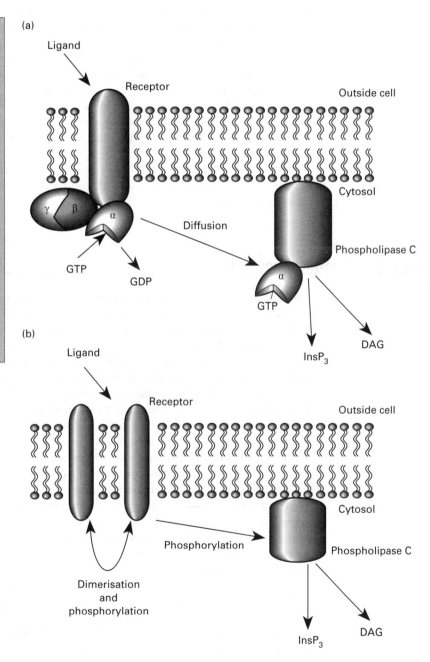

in the activation of this enzyme. When the cell is in the unstimulated state, PLCγ is mainly found in the cytosol. The tyrosine kinase-linked receptor not only phosphorylates PLCγ but also has to dimerise and phosphorylate itself. The phosphoryl groups added to the receptor itself create binding sites for the SH2 domains of PLCγ and so allow the translocation of PLCγ to the membrane and its association with the receptor. Therefore, such a mechanism brings PLCγ to the membrane where its substrate resides and, once associated with the receptor, phosphorylation activates the enzyme.

Interestingly, phosphorylation on a serine residue by both PKC and cAPK has been reported to be inhibitory, and therefore may reflect a negative control on the activity of PLC and the formation of intracellular signals.

As mentioned earlier, inositol-containing lipids have been found associated with the cytoskeleton. If the lipid is associated with profilin, an actin-binding protein, then the activity of PLCγ is severely reduced, suggesting that this too might be a regulatory mechanism.

A PLC has also been found that hydrolyses phosphatidylcholine (PC). This enzyme is a heterodimer of 69 and 55 kDa and preferentially hydrolyses PC rather than PtdInsP$_2$ as a substrate.

Inositol 1,4,5-trisphosphate and its Fate

Although InsP$_3$ was originally the only inositol phosphate formed from the hydrolysis of the inositol lipids to be assigned a role in cell signalling, it is clear that the hydrolysis of membrane lipids is only the first step in a very convoluted and complicated pathway. The hydrolysis of PtdIns leads to the formation of inositol 1-phosphate, while the hydrolysis of PtdIns 4-P leads to the formation of inositol 1,4-bisphosphate with InsP$_3$ itself arising from the hydrolysis of PtdInsP$_2$. However, inositol has the capacity to contain a variable number of phosphate groups, ranging from none, i.e. inositol, right through the spectrum of compounds to inositol hexaphosphate (InsP$_6$). InsP$_3$ is poised in the middle of this jungle of compounds and therefore can lead to the formation of lower inositol phosphates, i.e. forms containing two or one phosphate group, or to the formation of higher phosphate forms, i.e. four, five or six phosphate-containing compounds. Some, but not all, of the isomers and their interconversions are shown in Figure 6.5.

The addition of phosphate groups to the inositol ring is catalysed, as expected, by kinase enzymes, whereas phosphate removal is catalysed by phosphatases. This interconversion of the phosphate forms serves to deactivate certain inositol phosphates, for example lowering the concentration of InsP$_3$ and therefore reducing its capacity to cause the release of Ca^{2+} from intracellular stores, and so in effect turning off that part of the signal. However, InsP$_3$ is not the only inositol phosphate to have a signalling role. As reports appear in the literature, more and more of these inositol phosphates are being assigned roles in the control of cellular functions. Thus the removal of one form, reducing the signal associated with the rise in its concentration, may well lead to a rise in the concentration of another form and the formation

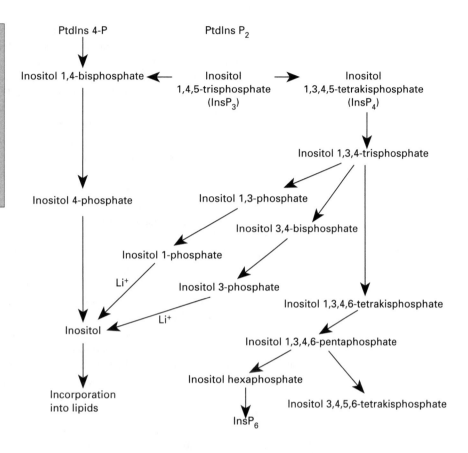

Figure 6.5
Some of the most important pathways of inositol phosphate metabolism, with the influence of lithium (Li⁺) ions included. PtdIns 4-P, Phosphatidylinositol 4-phosphate; PtdInsP₂, Phosphatidylinositol 4,5-bisphosphate.

of a new signal. It is beyond the scope of this book to attempt to look at all the isoforms involved and their postulated roles but some of the more prominent ones and the enzymes involved are discussed.

As mentioned, $InsP_3$ lies in the middle of the inositol phosphate phosphorylation states, where phosphates can be added or removed to form other compounds. Addition of a phosphate is carried out by $InsP_3$ 3-kinase, yielding the new inositol, $InsP_4$. The enzyme is soluble, uses ATP as a substrate and requires Mg^{2+}. The K_m for the substrate is lower than that of phosphatases that remove phosphate groups from $InsP_3$ and hence this might be the favoured reaction. The enzyme contains two 53-kDa catalytic subunits and also contains calmodulin, allowing stimulation via the Ca^{2+} pathway, while additional regulation of the kinase is almost certainly via phosphorylation, either by PKC or cAPK.

As for the role of $InsP_4$ some debate has occurred. It is thought that it is involved, like $InsP_3$, in the immobilisation of Ca^{2+} from internal stores, such as seen with histamine receptor stimulation, but some papers refute the role of $InsP_4$ in Ca^{2+} release. Some studies suggest that $InsP_4$ stimulates ATP-independent uptake of Ca^{2+} into the sarcoplasmic reticulum.

Certainly proteins that are capable of binding $InsP_4$ have been reported. One such protein, isolated from pig cerebellar tissue, has a molecular mass of 42 kDa and binds

$InsP_4$ 100 times better than $InsP_3$. Putative plasma membrane-associated receptors for $InsP_4$ in human platelets have been reported, whereas other groups have reported two $InsP_4$-binding sites associated with the nuclear envelope. One of these is in the outer nuclear membrane, a 74-kDa protein in rat liver, binds to $InsP_4$ with relatively high affinity and is involved in the uptake of nuclear Ca^{2+}. The other binding protein is on the inner membrane and has much lower $InsP_4$-binding affinity.

Interestingly, binding of both $InsP_3$ and $InsP_4$ to putative receptor sites in rat cortical membranes appears to be enhanced by peptides derived from β-amyloid protein. Therefore the production of β-amyloid may have drastic effects on Ca^{2+} signalling and such research may be relevant to the understanding of conditions such as Alzheimer's disease.

As well as the tetrakis forms of the inositol phosphates, further phosphate groups can be added to create the pentaphosphate and hexaphosphate derivatives. There are potentially six possible forms of inositol pentaphosphate, but in animals it appears that the most favoured form is inositol 1,3,4,5,6-pentaphosphate. However, most of the six isomers have been found in plants and in *Dictyostelium* suggesting that these too might have a defined role. The pentaphosphates, at least in animals, appear to be produced by the route outlined in Figure 6.4, i.e. $InsP_3$ is converted to $InsP_4$ which, by the removal of a phosphate from position 5, creates the trisphosphate, inositol 1,3,4-trisphosphate. Addition of a phosphate to position 6 leads to the production of inositol 1,3,4,6-tetrakisphosphate and further addition of a phosphate to position 5 produces inositol 1,3,4,5,6-pentaphosphate. Once produced, this new inositol compound can lead to the production of lower inositols, such as inositol 3,4,5,6–tetrakisphosphate or inositol 1,4,5,6-tetrakisphosphate, or by the further addition of a phosphate group, to the production of $InsP_6$. Inositol 3,4,5,6–tetrakisphosphate has been shown to accumulate in a dose–response related manner but its exact role remains to be determined, although it has been shown to selectively block epithelial Ca^{2+}-activated chloride channels. In the mean time, it has been dubbed an 'orphan signal'.

Upon stimulation of some cells, for example the promyeloid cell line HL-60, the concentrations of inositol pentaphosphate and inositol hexaphosphate rise rapidly, certainly within minutes, although in other systems the peak of accumulation of these compounds takes several hours if not days. However, the exact function of these higher inositol phosphates is unclear and it has been proposed that the main pentaphosphate form may just be a source of the four-phosphate derivative inositol 3,4,5,6-tetrakisphosphate. It has even been proposed that the pentaphosphate and hexaphosphate forms might have an extracellular role. Using cerebellar membranes, an $InsP_6$-binding protein has been found, but the exact functions of such proteins have yet to be discovered.

As well as the addition of phosphates to the inositol ring, their removal is also crucially important, not only to create more derivatives of the inositol phosphates but also to liberate free inositol, which is used to make new inositol lipids.

The removal of the phosphate from the 5 position of several inositol phosphates is carried out by inositol polyphosphate 5-phosphatase. This enzyme can use $InsP_3$ as a substrate, so creating inositol 1,4-bisphosphate. This inositol is also formed by the hydrolysis of PtdIns 4-P. Removal of the phosphate from position 5 of $InsP_3$ removes $InsP_3$ from the signalling pathway and so effectively turns off its message. However, this enzyme can also use InsP$_4$ and cyclic inositol 1:2,4,5-trisphosphate as substrates. The enzyme exists in many isoforms, which have been found located in both the soluble and particulate fractions of cells. In human platelets, for example, two immunologically distinct forms have been isolated, a 45-kDa Mg^{2+}-requiring enzyme (type I) and a 75-kDa enzyme (type II). The type I enzyme can be phosphorylated by PKC with a concomitant increase in its activity, suggesting that PKC may be instrumental in the reduction of InsP$_3$ levels in cells and act as a trigger for the termination of the InsP$_3$ signal. In other tissues other isoforms have been isolated, with diverse molecular masses, and in some cases substrate specificities vary.

Removal of the phosphate from the 5 position of InsP$_4$ leads to the production of another trisphosphate, inositol 1,3,4-trisphosphate. This molecule itself can undergo dephosphorylation to produce inositol 3,4-bisphosphate. The enzyme involved here is inositol polyphosphate 1-phosphatase. Its only other recorded substrate is another bisphosphate, inositol 1,4-bisphosphate and in this case inositol 4-phosphate is formed. The enzyme is monomeric with a molecular mass of approximately 44 kDa. Again there is a requirement for Mg^{2+}, but it is inhibited by the presence of Ca^{2+}. This reaction is also inhibited by lithium ions. They inhibit the activity of this enzyme and another phosphatase, inositol monophosphatase, and hence prevent the recycling of inositol, ultimately preventing its reincorporation into PtdIns lipids. Lithium ions have been used for years as a treatment for manic depression, and lithium ion concentrations found in patients undergoing this treatment are consistent with the K_i of the inhibition by lithium ions of these enzymes, adding weight to the hypothesis that it is through this action that lithium is having its therapeutic affect.

The recycling of inositol may also proceed via a different route. As well as the conversion of inositol 1,3,4-trisphosphate to the bisphosphate, inositol 3,4-bisphosphate, an enzyme called inositol polyphosphate 4-phosphatase can convert inositol 1,3,4-trisphosphate to a different bisphosphate, inositol 1,3-bisphosphate. It is also responsible for the conversion of inositol 3,4-bisphosphate to inositol 3-phosphate. This relatively large enzyme, of molecular mass 110 kDa, is monomeric and does not have a requirement for metal ions.

The inositol 1,3-bisphosphate formed is converted to inositol 1-phosphate by the enzyme inositol polyphosphate 3-phosphatase. This enzyme exists in two isoforms in rat brain. Electrophoresis has suggested that both exist as dimers, where type I appears as a dimer of 65-kDa subunits while the type II looks to be a heterodimer of 65 and 78-kDa subunits, with the 65-kDa subunits in the two cases probably being the same. The 78-kDa subunit of the type II isoform probably has a regulatory role as it seems to decrease the efficiency of catalysis. Interestingly, the substrate specificity of this enzyme is not restricted to soluble forms of the inositols and, amazingly, this enzyme can catalyse the removal of phosphate groups from the lipid PtdIns 3-phosphate to produce PtdIns. Therefore this enzyme might be involved in two disparate arms of the inositol signalling pathway.

Once the inositol phosphate has undergone dephosphorylation by the array of enzymes and has been reduced to the monophosphate form, the final dephosphorylation is carried out by the enzyme inositol monophosphatase. This enzyme removes phosphates from all positions around the inositol ring except those attached at the 2 position. Again Mg^{2+} is required for activity. The enzyme has an apparent molecular mass of 55 kDa and exists as a dimer of identical subunits. This enzyme is the second enzyme in the inositol phosphate pathway inhibited by lithium ions and, again, the lack of production of free non-phosphorylated inositol prevents its reincorporation into PtdIns and hence prevents further production of $InsP_3$ and continual cycling, and therefore signalling.

With the complete removal of all the phosphate groups, the resultant inositol is reused in the formation of PtdIns lipids in the endoplasmic reticulum, which are then reincorporated into the plasma membrane, ready for another round of signalling.

As well as the phosphorylated forms of inositol described above (see Figure 6.5) there are also cyclic forms (Figure 6.6). If cells are stimulated for long periods of time,

Figure 6.6
Molecular structure of a cyclic inositol compound.

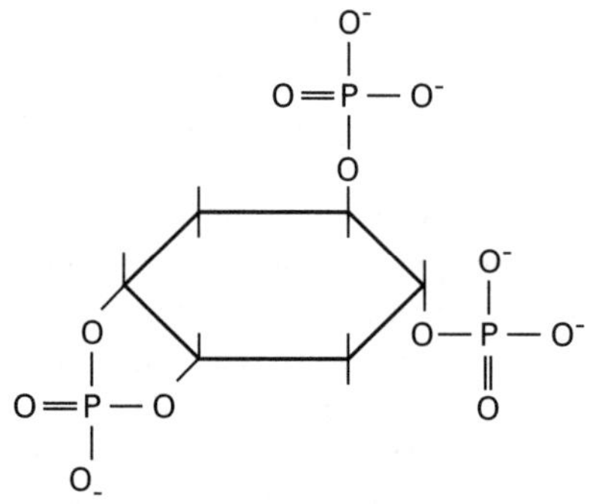

the cyclic forms of the inositol phosphates tend to increase. Like their non-cyclic counterparts, they are produced by the action of PLC but they are in general not substrates for the enzymes that add and remove phosphates from the non-cyclic forms of inositol phosphates. In the normal PLC reaction, hydrolysis of the bond between the phosphate of the inositol ring and the glycerol backbone of the lipid involves the supply of a hydroxyl group from water. However, this hydroxyl group can also be supplied from the inositol ring itself, and if this happens then the phosphate is cyclised. The only enzyme reported to be able to break the cyclic phosphate bonds is inositol 1:2-phosphate 2-phosphohydrolase or cyclic hydrolase. However, the only substrate for this enzyme is inositol cyclic 1:2-phosphate. Therefore, it is presumed that all the cyclic inositol phosphates are finally metabolised through this route and therefore their concentrations in the cell are probably controlled by this enzyme, along with the rate of their production by PLC. The activity of cyclic hydrolase is inhibited by the metal ion Zn^{2+} but activated by Mn^{2+}. Inhibition also occurs with inositol 2-phosphate. Interestingly, when cyclic hydrolase was isolated from human placenta it was found to be the same as another protein, lipocortin III. This protein was known to bind lipids and Ca^{2+}, and belonged to a group of eight related proteins. By the use of cDNA and overexpression techniques, the activity of cyclic hydrolase has been correlated with levels of proliferation of cell culture. It has been suggested that the enzyme is antiproliferative and could be an example of an anti-oncogene, although cyclic hydrolase is not able to prevent transformation of cells.

Other forms of inositol phosphates are the inositol pyrophosphates, where two phosphates are joined together and then attached to the inositol ring. Such compounds have been mainly found in *Dictyostelium*. The pyrophosphate groups are usually at the 1 or 3 positions of the ring but pyrophosphate attached to the 4 and 6 positions have also been seen. Similarly, mammalian cells contain pyrophosphate inositols formed by ATP-dependent phosphorylation of inositol 1,3,4,5,6-pentaphosphate and $InsP_6$ (designated $InsP_5P$ and $InsP_6P$ respectively). However, once again, the significance of these findings has yet to be determined.

Role of Diacylglycerol

Any discussion of the role of DAG must emphasise that DAG is in fact not a single chemical but a family of related compounds, the structure of which is determined by the acyl groups present in the original lipid that was hydrolysed by the phospholipase. The main role of DAG has always been determined as an activator of PKC and there is clear evidence that activation of PLC leads to production of $InsP_3$ and DAG and this is accompanied by a rise in PKC activity.

However, there is a growing view that DAG may be further metabolised to other signalling compounds. One such compound is PA. This chemical stimulates inositol 4,5-bisphosphate formation, activates PLC and acts as a cell mitogen. Its production involves the phosphorylation of DAG by diacylglycerol kinase (DGK). Isoforms of this enzyme have been found in both the membrane and cytoplasmic fractions of mammalian cells, with the molecular masses of reported enzymes varying enormously. For example, the enzyme from porcine brain tissue appears to be 80 kDa, that from porcine thymus tissue is 80 and 150 kDa, while human platelets seem to have isoforms at 58, 75 and 152 kDa. Interestingly, when two of the isoforms with a molecular mass of approximately 80 kDa were cloned they were found to contain protein-folding regions known as zinc finger domains, suggesting binding and possible control by Ca^{2+} ions. Also receptor activation leads to regulation of DGK activity, while some forms are phosphorylated by PKC and cAPK.

The breakdown of DAG and hence the termination of its signal is an open question, but it appears that the exact route for its metabolism depends on the acyl groups involved.

Inositol Phosphate Metabolism at the Nucleus

Classically, the metabolism of inositol lipids has always been thought of as a plasma-membrane event. However, the same biochemical pathways also occur in the nucleus, with $PtdInsP_2$ and possibly PtdIns 4-P being broken down to $InsP_3$ and DAG. Interestingly, many of the studies involving the nucleus have been carried out after the removal of the nuclear membrane, suggesting that it is the lipids and enzymes that are associated with skeletal structures within the nucleus that are involved and not the lipids that constitute the nuclear membrane. This has obvious analogies to the association of inositol lipids with the cytoplasmic cytoskeleton. Activation of this nuclear pathway is probably involved in the control of the cell cycle, with the likely involvement of PKC activated by the released DAG. DNA synthesis may also be under the control of inositol derivatives such as PtdIns 4-P or inositol 1,4-bisphosphate.

Other Lipids Involved in Signalling

Phosphatidylcholine and Arachidonic Acid Metabolism

The phosphatidylinositols are not the only lipids to be involved in signalling from the plasma membrane to the inside of the cell. It is well documented that PC is also the precursor of signalling molecules. PC can be hydrolysed by several enzymes: phospholipase A_2 (PLA$_2$), PLC, or PLD (Figure 6.7).

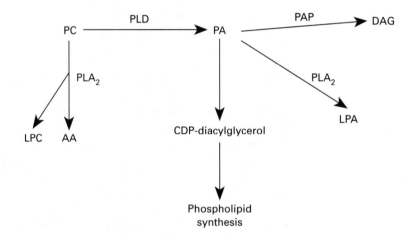

Figure 6.7
The breakdown of phosphatidylcholine (PC) by phospholipase A$_2$ (PLA$_2$) and phospholipase D (PLD). AA, arachidonic acid; DAG, diacylglycerol; LPA, lysophosphatidic acid; LPC, lysophosphatidylcholine; PA, phosphatidic acid; PAP, phosphatidate phosphohydrolase.

As a general rule, agonists that lead to the hydrolysis of PC also cause an increase in the hydrolysis of PtdIns in the same cell. However, the rise in DAG is biphasic. Firstly, the breakdown of PtdIns leads to a rapid increase in the concentration of DAG; slightly later, the hydrolysis of PC leads to a more prolonged accumulation of DAG, along with the production of choline and phosphorylcholine.

However, PC, along with the breakdown of PtdIns and phosphatidylethanolamine (PE), is the major source of intracellular arachidonic acid (AA), which itself has been implicated as a major signalling molecule as well as leading to the production of eicosanoids. As AA is most commonly found in the second or middle position of the glycerol backbone of the phospholipid, its release is catalysed by the enzyme PLA$_2$ (Figure 6.8). PLA$_2$ is associated with the plasma membrane of cells, but can also be found in the cytoplasm and associated with intracellular membranes. Purification has

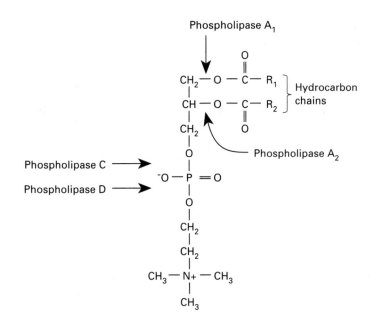

Figure 6.8
Molecular structure of phosphatidylcholine and its breakdown by phospholipase A$_2$, phospholipase A$_1$, phospholipase D and phospholipase C. R denotes the fatty acid chains, where R$_2$ is likely to be arachidonic acid.

shown that the enzyme is usually about 80–100 kDa in mass and shows activation by Ca^{2+}, with concomitant translocation to the membrane. The N-terminal region of the enzyme contains a Ca^{2+}-dependent phospholipid-binding sequence that mediates its interaction with the membrane. PLA$_2$ also hydrolyses PtdIns and PE at the second carbon of the glycerol backbone, and although AA is the usual fatty acid at this position, PLA$_2$ is not specific for this substrate.

PLA$_2$ enzymes are also extracellular, for example in snake venom, synovial fluid and pancreatic secretions. These enzymes have a very low molecular mass less than 20 kDa, and fall into one of two groups, PLA$_2$-I or PLA$_2$-II, which are generally not thought to be involved in cell-signalling cascades.

The activation of PLA$_2$ activity is certain to be complex. Besides its activation by Ca^{2+}, it has also been reported that PLA$_2$ is activated by phosphorylation by MAPK as well as by specific isoforms of PKC. PKC can also lead to the activation of the MAPKs that cause the phosphorylation and activation of PLA$_2$. As well as the phosphorylation of the enzyme, it is thought that G proteins also have a role in PLA$_2$ regulation. Although effects have been seen with addition of GTPγ-S, with locks G proteins in their active state, direct interaction between a G protein and the enzyme has to be demonstrated and it is possible that G proteins are themselves having an indirect effect through a kinase.

AA produced by the hydrolysis of PC by PLA$_2$ is a 20-carbon unsaturated fatty acid containing four double bonds. It can itself act as a signal, but it can also lead to the production of prostaglandins, thromboxanes, leukotrienes and other eicosanoids. Some of the enzymes that act on AA or are involved in its further metabolism include cyclooxygenase, cytochrome P450 and lipoxygenases (Figure 6.9). Cyclooxygenase, otherwise known as prostaglandin G/H synthase (PGHS), leads to

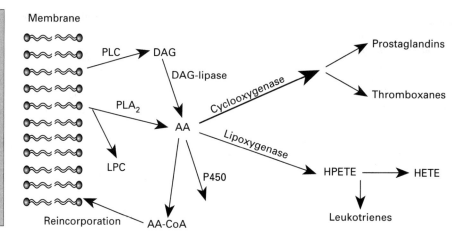

Figure 6.9
Central role of arachidonic acid (AA): its production and further metabolism. DAG, diacylglycerol; HETE, hydroxyeicosatetraenoic acid; HPETE, hydroperoxyeicosatetraenoic acid LPC, lysophosphatidylcholine; PLA$_2$, phospholipase A$_2$, PLC, phospholipase C.

the production of prostaglandins G$_2$ and H$_2$ (PGG$_2$ and PGH$_2$), which further leads to the formation of other prostaglandins, prostacyclins and thromboxanes. Cytochrome P450 leads to the production of epoxyeicosatrienoic acids (EET), which can be acted on by epoxide hydrolase to release diols. Lipoxygenases convert AA to hydroperoxyeicosatetraenoic acids (HPETE) leading to the formation of hydroxyeicosatetraenoic acids (HETE). Alternatively HPETE can lead to the production of leukotrienes and epoxyhydroxides. Added confusion arises when it is realised that some of these enzymes, for example lipoxygenases, can utilise other polyunsaturated fatty acids as well as AA, for example linoleic acid, that are released by PLA$_2$, and this may also lead to the production of even more molecules with signalling potential. Furthermore, AA can arise from the hydrolysis of DAG via the action of DAG lipase. DAG, as discussed, is produced from the hydrolysis of phospholipids by PLC, and normally would be used in the activation of PKC. However, through the action of DAG lipase, DAG also leads to AA and its further metabolism.

Many of the products of AA metabolism have regulatory roles within the cell. Ion channels, Na$^+$/K$^+$ ATPase activity and cell proliferation are amongst a whole host of functions under such control. Receptors for some of the eicosanoid family have recently been cloned. These, not surprisingly, contain seven putative membrane-spanning regions and are thought to act through trimeric G proteins in the cell. As well as its further products, AA itself can lead to the activation of some isoforms of PKC and to the activation of PLD. No doubt, more and more regulatory functions for AA and the eicosanoids will come to light.

AA can also be used in the re-formation of membrane lipids. This is through a route that includes, firstly, its conversion to arachidonoyl-coenzyme A by the enzyme arachidonoyl CoA synthetase and, secondly, an esterification catalysed by the enzyme arachidonoyl-lysophospholipid transferase. This route obviously reduces the AA concentration in the cell available for the production of the host of signalling molecules previously mentioned, as well as reduces the signal arising directly from the presence of the AA.

The formation of AA is also modulated by the presence of a 37 kDa protein called lipocortin, otherwise known as annexin, possibly through the inhibition of PLA$_2$. The identity of lipocortin was first discovered by studies of the suppression of eicosanoid production by glucocorticoid treatment and is in fact a family of at least 12 proteins. Different members of the family are characterised by their variable N-terminal regions, but all lipocortins contain between four and eight highly conserved repeating domains. These domains are capable of binding to phospholipids in the presence of Ca^{2+} ions. However, studies including the use of peptides designed from the known sequences of the lipocortins have revealed that the proteins may have far wider roles in cellular regulation, and their presence has been implicated in the control of neutrophil migration, differentiation and growth of cells, and in the protection of neuronal tissues.

Hydrolysis of PC by PLA_2 not only produces AA but leaves behind a by-product, lysophosphatidylcholine (LPC). This too is thought to have a signalling role, and although its exact action has not been fully defined it probably acts through PKC.

Another enzyme with an important role in the hydrolysis of PC is PLD (see Figure 6.8). This cleavage of the phospholipid yields PA as a product, and it is thought that a significant amount of DAG in a cell arises from the further breakdown of PA by the enzyme phosphatidate phosphohydrolase (PAP). Of course this is contrary to the production of PA from DAG by DAG kinase as discussed earlier. This breakdown of PA to DAG, however, is increased by several agonists in a wide range of cells. PAP exists in at least two forms, which differ in their subcellular locations as well as their enzymatic properties. PLD has not been fully purified but preliminary studies have shown that it has a likely molecular mass of approximately 200 kDa and, interestingly, appears to exist in several isoforms with varying subcellular locations.

Early work was based on the observation that PLD activity is modulated by the addition of stable non-hydrolysable GTP analogues such as GTPγ-S. However, the factor shown to be the activator of PLD activity is a low-molecular-mass GTP-binding protein known as ARF. This protein was originally found to be required for the ADP-ribosylation of the G protein subunit $G_{s\alpha}$ in the presence of cholera toxin. ARF has also been shown to have a role in the control of vesicle protein trafficking. Other groups have suggested the involvement of another G protein, Rho, in the regulation of PLD, while further control of the activities of PLD and PLC in the hydrolysis of PC comes from PKC. As well as serine/threonine phosphorylation, tyrosine phosphorylation has been shown to be involved here. Interestingly, hydrogen peroxide causes an increase in PLD activity; this effect could be mediated by a tyrosine phosphorylation step, although evidence of a direct tyrosine phosphorylation of the PLD polypeptide leaves the exact mechanism of activation open to debate.

The main role of PLD-catalysed breakdown of lipids may be to increase DAG concentrations in the cell, but the product itself (PA) may also have signalling properties. It is thought that PA may have regulatory roles on other signalling molecules such as adenylyl cyclase, kinases and phosphatases. PA has also been implicated in the control of superoxide ion release neutrophils and in thrombin-induced actin polymerisation of fibroblasts.

The action of PLA_2 on PA leads to the production of LPA. This molecule has been shown to be an extracellular signal. Receptors for LPA have been found on some cells by the use of photoreactive analogues of LPA although, once again, much work is needed to elucidate fully the exact mechanisms involved.

As mentioned above, the concentration of DAG in the nucleus is of crucial importance for the regulation of many nuclear events, and a substantial proportion of this DAG may be a result of the breakdown of PC.

An important point that should be remembered with any hydrolysis of membrane lipids, be it PtdIns, PC or PE, is that lipids are being removed from the membrane and therefore the structure and integrity of that membrane may be altered. This itself may well be acting as a control pathway, besides the obvious effects of the products being made. The alteration of lipid membrane fluidity caused by the removal of certain lipids may be involved in the regulation of secretion, neurotransmission and in the control of the activity of some membrane proteins. For example, detergents have been shown to activate superoxide ion release from neutrophils, the enzyme involved probably responding to alteration of the lipid structures of the membrane.

Sphingolipid Pathways

A third pathway involving the breakdown of lipids, independent of the PLC, PtdIns and PC pathways, has come to light in recent years. Sphingosine was reported to inhibit PKC activity, which led to a flurry of research that revealed what is referred to as the sphingomyelin cycle. Essentially, extracellular signals such as IFN-γ, IL-1 and complement lead to the activation of an enzyme, sphingomyelinase, which hydrolyses sphingomyelin (SM) releasing ceramide.

Unlike other phospholipids, SM does not have a backbone of glycerol but rather of sphingosine (Figure 6.10). Sphingosine is an amino alcohol that contains a long unsaturated hydrocarbon chain. To produce SM, a fatty acid is added to sphingosine by an amide bond, creating ceramide, while its primary hydroxyl group is esterified to phosphorylcholine. In fact, in the biosynthesis, SM is made by an exchange of the phosphorylcholine head group from PC to ceramide, catalysed by the enzyme phosphatidylcholine:ceramide choline phosphotransferase. Breakdown by sphingomyelinase once again yields ceramide, which can enter a catabolic cascade.

Figure 6.10
Molecular structure of sphingomyelin showing its backbone of sphingosine.

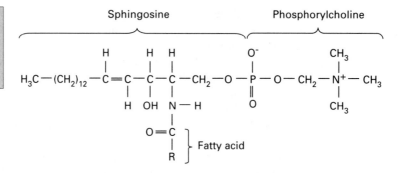

Ceramide is an important regulator in protein trafficking, cell growth, differentiation and in the viability of cells, having an important role in the orchestration of apoptosis. On a biochemical level, ceramide activates a nuclear factor, NF-κB, in permeabilised cells and is important in the control of transcription of several genes. Ceramide also has an effect through phosphorylation, with activation of MAPK. However, it has been shown to have a major control route through the use of a phosphatase. A serine/threonine phosphatase has been identified that is specifically and directly activated by the presence of ceramide and this enzyme has been called ceramide–activated protein phosphatase (CAPP; Figure 6.11). This phosphatase

Figure 6.11
Production of ceramide and one of its possible roles within the cell.

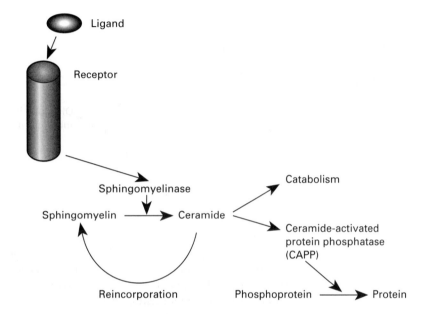

belongs to the 2A group of serine/threonine protein phosphatases and therefore has a trimeric structure. This seems like an elegant signalling cascade with ceramide as a central player, the overall result being opposite to the promotion of protein phosphorylation seen with activation of PKC by DAG.

Sphingosine itself stimulates the hydrolysis of PtdIns, resulting in the production of InsP$_3$ and the subsequent release of Ca^{2+} ions. However, another metabolite, sphingosine 1-phosphate, has second-messenger activity and is involved in the inositol-independent release of Ca^{2+} ions from intracellular stores (discussed in Chapter 7).

Related Lipid-derived Signalling Molecules

It is now evident that many of the developmental stages of plant growth, as well as many cellular functions of plants, are under the control of complex compounds which can be derived from lipids. Some of these compounds were first shown to be of interest as they could induce tuber formation in plants such as potatoes and yams. One of the major compounds is jasmonic acid (JA). JA is produced by a short metabolic pathway that starts by the action of lipoxygenase, which catalyses the incorporation of O_2 into linolenic acid. Most of the linolenic acid in the cell is found associated with lipids of the plasma membrane and can be released by the action of PLA_2. The 13-hydroperoxylinolenic acid produced by the action of PLA_2 is further acted upon sequentially by hydroperoxide dehydrase, allene oxide cyclase and a reductase. Three further β-oxidation steps finally yield the 12-carbon JA (Figure 6.12).

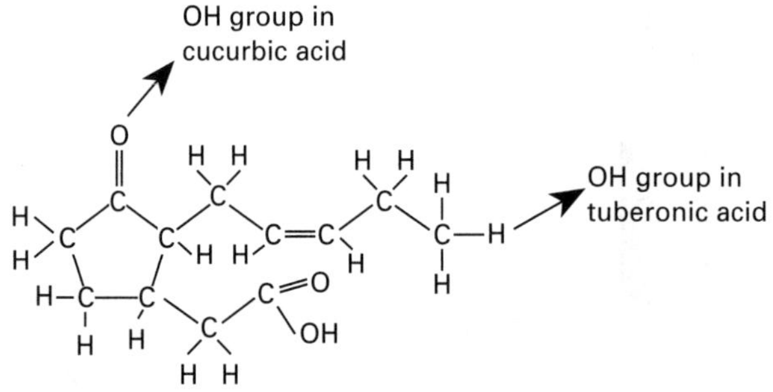

Figure 6.12
Molecular structure of jasmonic acid with derivatives shown.

Other derivatives of this compound include methyl JA, cucurbic acid and tuberonic acid, the latter named after its tuber-inducing capacity. Just as with the inositol pathway, further derivatives of these compounds have been isolated and, because of the presence of chiral carbon atoms, several stereoisomers can exist, although there is probably a favoured conformation that occurs naturally.

JA and methyl JA induce senescence in leaves and promote tuber formation. If leaves were floated on a solution of JA for 4 days, it was found that the polypeptide profile of the leaves changed as analysed by SDS-PAGE. These new proteins were termed jasmonate-induced proteins (JIPs). Further analysis of the regulation of the synthesis of these proteins indicated that control was probably at the levels of transcription and translation. JA also induces the expression of other proteins with a known specific function, including storage proteins of leaves, known as vegetative storage protein (VSP). These proteins accumulate in vacuoles of the mesophyllic cells of the leaf and the bundle sheath cells before flowering. The mRNA coding for VSP is increased on the addition of JA. Cell division has also been suggested as a target for control by JA or its related compounds.

Summary

One of the most central parts of many cell-signalling pathways is the breakdown of lipids in the plasma membrane to form messenger molecules that transmit the signal into the cell. Recently, the formation of new inositol lipids by PtdIns 3-kinase has been shown to be important in many transduction events, although the exact action of the lipids remains obscure. More fully understood is the breakdown of $PtdInsP_2$ by PLC, with the formation of $InsP_3$ and DAG. DAG is known to act as an activator of PKC while $InsP_3$ releases Ca^{2+} from intracellular stores (see Figure 6.1). PLC may be activated by the trimeric G protein G_q. The complexity of the system can be seen when it is realised that $InsP_3$ may be the precursor of both higher and lower inositol phosphates, i.e. it can be phosphorylated or dephosphorylated to create new metabolites, each of which may have a signalling role within the cell. Several of these inositol phosphates are thought to have defined roles, including $InsP_4$, and binding proteins for several of these compounds have now been reported. Several of the enzymes involved in this metabolism are the target for lithium ion inhibition, which might account for its therapeutic action in treating manic depression. Other inositols, such as cyclic phosphate isomers and pyrophosphates, have also been reported.

Other lipids too can lead to the formation of signalling molecules. Here, lipid-hydrolysing enzymes such as PLD and PLA_2 are involved. One of the most important metabolites is AA, which can be the substrate for cyclooxygenase, cytochrome P450 and lipoxygenase, leading to the formation of prostaglandins, leukotrienes and thromboxanes.

Other lipid metabolites that have a clear influence on cellular functioning are those involving the sphingosine metabolites such as ceramide and sphingosine 1-phosphate, while in plants JA has a diverse range of actions, including the induction of senescence of leaves and tuber formation.

Further Reading

Berridge, M. J. (1987) Inositol trisphosphate and diacylglycerol: two interacting second messengers. *Annual Review of Biochemistry*, **56**, 159–93.

Berridge, M. J. (1993) Inositol trisphosphate and calcium signalling. *Nature*, **361**, 315–26.

Majerus, P. W. (1992) Inositol phosphate biochemistry. *Annual Review of Biochemistry*, **61**, 225–50.

Events at the Membrane

Agranoff, B. W., Bradley, R. M. and Brady, R. O. (1958) The enzymatic synthesis of inositol phosphatide. *Journal of Biological Chemistry*, **233**, 1077–83.

Bansal, V. S. and Majerus, P. W. (1990) Phosphatidylinositol derived precursors and signals. *Annual Review of Cell Biology*, **6**, 41–67.

Carpenter, C. L. and Cantley, L. C. (1990) Phosphoinositide kinases. *Biochemistry*, **29**, 11147–56.

Divecha, N. and Irvine, R. F. (1995) Phospholipid signalling. *Cell*, **80**, 269–78.

Goldschmidt-Clermont, P. J., Kim, J. W., Machesky, L. M., Rhee, S. G. and Pollard, T. D. (1991) Regulation of phospholipase C-γ1 by profilin and tyrosine phosphorylation. *Science*, **251**, 1231–3.

Janmey, P. A. (1994) Phosphoinositides and calcium as regulators of cellular actin assembly and disassembly. *Annual Review of Physiology*, **56**, 169–91.

Kodaki, T., Worchoski, R., Hallberg, R., Rodriguez-Viciana, P., Downward, J. and Parker, P. J. (1994) The activation of phosphatidylinositol 3-kinase by Ras. *Current Biology*, **4**, 798–806.

Lassing, I. and Lindberg, V. (1988) Evidence that the phosphatidylinositol cycle is linked to cell mobility. *Experimental Cell Research*, **174**, 1–15.

Ling, L. E., Schulz, J. T., Cantley, L. C. (1989) Characterisation and purification of membrane associated phosphatidylinositol-4-phosphate kinase from human red blood cells. *Journal of Biological Chemistry*, **264**, 5080–8.

Majerus, P. W., Neufeld, E. J. and Wilson, D. B. (1984) Production of phospho-inositide derived messengers. *Cell*, **37**, 701–3.

Majerus, P. W., Connolly, T. M., Deckmyn, H. *et al.* (1986) The metabolism of phosphoinositol derived messenger molecules. *Science*, **234**, 1519–26.

Majerus, P. W., Ross, T. S., Cunningham, T. W., Caldwell, K.K ., Jefferson, A. B. and Bansal, V. S. (1990) Recent insights in phosphatidylinositol signaling. *Cell*, **63**, 459–65.

Nishibe, S., Wahl, M. I., Rhee, S. G. and Carpenter, G. (1989) Tyrosine phospho-rylation of phopholipase C-II *in vitro* by the epidermal growth factor receptor. *Journal of Biological Chemistry*, **264**, 10335–8.

Okada, T., Sakuma, L., Fukui, Y., Hazeki, O. and Ui, M. (1994) Blockage of chemotactic peptide induced stimulation of neutrophils by Wortmannin as a result of selective inhibition of phosphatidylinositol 3-kinase. *Journal of Biological Chemistry*, **269**, 3563–3567.

Putney, J. W., Takemura, H., Hughes, A. R., Horstman, D. A. and Thastrup, O. (1989) How do inositol phosphates regulate calcium signalling? *FASEB Journal*, **3**, 1899–905.

Rodriguez-Viciana, P., Warne, P .H., Dhand, R. *et al.* (1994) Phosphatidylinositol-3-OH kinase as a direct target of Ras. *Nature*, **370**, 527–32.

Smrcka, A. V., Hepler, J. R., Brown, K. O. and Sternweis, P. C. (1991) Regulation of polyphosphoinositide specific phospholipase C activity by purified G_q. *Science*, **251**, 804–7.

Wennstrom, S., Hawkins, P. T., Cooke, F. *et al.* (1994) Activation of phosphoinosi-tide 3-kinase is required for PDGF-stimulated membrane ruffling. *Current Biology*, **4**, 385–93.

Whitman, M., Downes, C. P., Keeler, M., Keller, T. and Cantley, L. (1988) Type I phophatidylinositol kinase makes a novel inositol phospholipid, phosphatidylinos-itol 3-phosphate. *Nature*, **332**, 644–6.

Inositol 1,4,5-trisphosphate and its Fate

Bansal, V. S., Inhorn, R. C. and Majerus, P. W. (1987) The metabolism of inositol 1,3,4,-trisphosphate to inositol 1,3-bisphosphate. *Journal of Biological Chemistry*, **262**, 9444–7.

Bansal, V. S., Caldwell, K. K. and Majerus, P. W. (1990) The isolation and characterisation of inositol polyphosphate 4-phosphatases. *Journal of Biological Chemistry*, **265**, 1806–11.

Berridge, M. J., Downes, C. P. and Hanley, M. R. (1989) Neural and developmental actions of lithium: a unifying hypothesis. *Cell*, **59**, 411–19.

Caldwell, K. K., Lips, D. L., Bansal, V. S. and Majerus, P. W. (1991) Isolation and characterisation of 2,3,-phosphatases that hydrolyse the phosphatidylinositol 3-phosphate and inositol 1,3-bisphosphate. *Journal of Biological Chemistry*, **266**, 18378–86.

Connolly, T. M., Bross, T. E. and Majerus, P. W. (1985) Isolation of a phosphomonoesterase from human platelets that specifically hydrolyses the 5-phosphate of inositol 1,4,5-trisphosphate. *Journal of Biological Chemistry*, **260**, 7868–74.

Connolly, T. M., Lawing, W. J. and Majerus, P. W. (1986) Protein kinase C phosphorylates human platelet inositol trisphosphate 5' phosphomonoesterase, increasing the phosphatase activity. *Cell*, **46**, 951–8.

Cowburn, R. F., Wiehager, B. and Sundstrom, E. (1995) Beta-amyloid peptides enhance binding of the calcium mobilising second messengers, inositol (1,4,5)trisphosphate and inositol (1,3,4,5)tetrakisphosphate to their receptor sites in rat cortical membranes. *Neuroscience Letters*, **191**, 31–4.

Cullen, P. J., Patel, Y., Kakkar, V. V., Irvine, R. F. and Authi, K. S. (1994) Specific binding-sites for inositol 1,3,4,5-tetrakisphosphate are located predominantly in the plasma membranes of human platelets. *Biochemical Journal*, **298**, 739–42.

Donie, F. and Reiser, G. (1991) Purification of a high affinity inositol 1,3,4,5-tetrakisphosphate receptor from brain. *Biochemical Journal*, **267**, 453–7.

Gee, N. S., Reid, G. G., Jackson, R. G., Barnaby, R. J. and Ragan, C. I. (1988) Purification and properties of inositol 1,4-bisphosphate. *Biochemical Journal*, **253**, 777–82.

Hawkins, P. T., Reynolds, D. J., Poyner, D. R. and Hanley, M. R. (1990) Identification of a novel inositol phosphate recognition site: specific [H-3] inositol hexakisphosphate binding to brain regions and cerebellar membranes. *Biochemical and Biophysical Research Communications*, **167**, 819–27.

Inhorn, R. C. and Majerus, P. W. (1988) Properties of inositol polyphosphate 1-phosphatase. *Journal of Biological Chemistry*, **263**, 14559–65.

Inhorn, R. C., Bansal, V. S. and Majerus, P. W. (1987) Pathway for inositol 1,3,4–trisphosphate and 1,4-bisphosphate metabolism. *Proceedings of the National Academy of Science USA*, **84**, 2170–14.

Irvine, R. F. and Moore M. (1986) Microinjection of inositol 1,3,4,5-tetrakisphosphate activated sea urchin eggs by a mechanism dependent on external Ca^{2+}. *Biochemical Journal*, **240**, 917–20.

Irvine, R. F., Moor, R. M., Pollock, W. K., Smith, P. M. and Wreggett, K. A. (1988) Inositol phosphate: proliferation, metabolism and function. *Philosophical Transactions of the Royal Society of London Series B*, **320**, 281–98.

Ismailov, I. I., Fuller, C. M., Berdiev, B. K., Shlyonsky, V. G., Benos, D. J. and Barrett, K.E. (1996) A biological function for an orphan messenger: D-myo-inositol 3,4,5,6-tetrakisphosphate selectively blocks epithelial calcium activated chloride channels. *Proceedings of the National Academy of Science USA*, **93**, 10505–9.

Jackson, R. G., Gee, N. S. and Ragan C. I. (1989) Modification of myoinositol monophosphate by the arginine specific reagent phenylglyoxal. *Biochemical Journal*, **264**, 419–22.

Malviya, A. N. (1994) The nuclear inositol 1,4,5-trisphosphate and inositol 1,3,4,5-tetrakisphosphate receptors. *Cell Calcium*, **16**, 301–13.

Menniti, F. S., Oliver, K. G., Putney, J. W. and Shears, S. B. (1993) Inositol phosphates and cell signalling: new views of InsP$_5$ and InsP$_6$. *Trends in Biochemical Science*, **18**, 53–6.

Mitchell, C. A., Connolly, T. M. and Majerus, P. W. (1989) Identification and isolation of a 75kDa inositol polyphosphate 5-phosphatase from human platelets. *Journal of Biological Chemistry*, **264**, 8873–7.

Nicholetti, F., Bruno, V., Cavallar, S., Copani, A., Sortino, M. A. and Canono, P. L. (1990) Specific binding-sites for inositol hexakisphosphate in brain and anterior pituitary. *Molecular Pharmacology*, **37**, 689–693.

Quist, E. E., Foresman, B. H., Vasan, R. and Quist, C. W. (1994) Inositol tetrakisphosphate stimulates a novel ATP-independent Ca^{2+} uptake mechanism in cardiac junctional sarcoplasmic reticulum. *Biochemical and Biophysical Research Communications*, **204**, 69–75.

Ross, T. S. and Majerus, P. W. (1986) Isolation of D-myo-inositol 1-2-cyclic phosphate 2-inositolphosphohydrolase from human platelets. *Journal of Biological Chemistry*, **261**, 11119–23.

Ross, T. S. and Majerus, P. W. (1991) Inositol 1,2-cyclic phosphate 2-inositol phosphohydrolase: substrate specificity and regulation of activity by phospholipids, metal ion chelators and inositol 2-phosphate. *Journal of Biological Chemistry*, **266**, 851–6.

Ryu, S. H., Lee, S. Y., Lee, K. Y. and Rhee, S. G. (1987) Catalytic properties of inositol trisphosphate kinase: activation by Ca^{2+} and calmodulin. *FASEB Journal*, **1**, 388–93.

Stephens, L. R., Hawkins, P. T., Stanley, A. F. *et al.* (1991) Myoinositol pentakisphosphate: structure, biological occurrence and phosphorylation to myoinositol hexakisphosphate. *Biochemical Journal*, **275**, 485–99.

Takimoto, K., Okada, M., Matsuda, Y. and Nakagawa, H. (1985) Purification and properties of myoinositol 1-phosphate from rat brain. *Journal of Biochemistry*, **98**, 363–70.

Vallejo, M., Jackson, P., Lightman, S. and Hanley, M. R. (1987) Occurrence and extracellular actions of inositol pentakisphosphate and hexakisphosphate in mammalian brain. *Nature*, **330**, 656–8.

Role of Diacylglycerol

Florin-Christersen, J., Florin-Christersen, M., Delfino, J. M. and Rasmusssen, H. (1993) New pattern of diacylglycerol metabolism in intact cells. *Biochemical Journal*, **289**, 783–8.

Fujikawa, K., Imai, S., Sakane, F. and Kanoh., H. (1993) Isolation and characterisation of the human diacylglycerol kinase gene. *Biochemical Journal*, **294**, 443–9.

Ohanian, J. and Heagerty, A. M. (1994) Membrane-associated diacylglycerol kinase activity is increased by noradrenaline, but not by angiotensin II, in arterial smooth muscle. *Biochemical Journal*, **300**, 51–6.

Inositol Phosphate Metabolism at the Nucleus

Banfic, H., Zizak, M., Divecha, N. and Irvine, R. (1993) Nuclear diacylglycerol is increased during cell proliferation *in vivo*. *Biochemical Journal*, **290**, 633–6.

Divecha, N., Banfic, H. and Irvine, R. F. (1993) Inosities and the nucleus and inosities in the nucleus. *Cell*, **74**, 405–7.

Sylvia, V., Curtin, G., Norman, J., Stec, J. and Busbee, D. (1988) Activation of a low specific activity form of DNA polymerase α by inositol-1,4-bisphosphate. *Cell*, **54**, 651–8.

York, J. D., Saffritz, J. E. and Majerus, P. W. (1994) Inositol phosphate 1-phosphatase is present in the nucleus and inhibits DNA synthesis. *Journal of Biological Chenistry*, **259**, 19992–9.

Phosphatidylcholine and Arachidonic Acid Metabolism

Billah, M. M. and Anthes, J. C. (1990) The regulation and cellular functions of phosphatidylcholine hydrolysis. *Biochemical Journal*, **269**, 281–91.

Bowman, E. P., Uhlinger, D. J. and Lambeth, J. D. (1993) Neutrophil phospholipase D is activated by a membrane associated Rho family small molecular weight GTP binding protein. *Journal of Bological Chemistry*, **268**, 21509–12.

Brown, H. A., Gutowski, S., Moonaw, C. R., Slaughter, C. and Sternweis, P. C. (1993) ADP-ribosylation factor, a small GTP-dependent regulating protein stimulates phospholipase D activity. *Cell*, **75**, 1137–44.

Clark, J. D., Lin, L.-L., Kritz, R. W. *et al.* (1991) A novel arachidonic acid-selective cytosolic PLA$_2$ contains a Ca^{2+}-dependent translocation domain with homology to PKC and GAP. *Cell*, **65**, 1043–51.

Cockcroft, S., Thomas, G. M. H., Fensome, A. *et al.* (1994) Phospholipase D: a downstream effector of ARF in granulocytes. *Science*, **263**, 523–6.

Durieux, M. E. and Lynch, K. R. (1993) Signalling properties of lysophosphatidic acid. *Trends in Pharmacological Sciences*, **14**, 249–54.

Exton, J. H. (1990) Signalling through phosphatidylcholine breakdown. *Journal of Biological Chemistry*, **265**, 1–4.

Exton, J. H. (1994) Phosphoinositol phospholipases and G proteins in hormone actions. *Annual Review of Physiology*, **56**, 349–69.

Exton, J. H. (1994) Phosphatidylcholine breakdown and signal transduction. *Biochimica et Biophysica Acta*, **1212**, 26–42.

Flower, R. J. and Rothwell, N. J. (1994) Lipocortin-1: cellular mechanisms and clinical relevance. *Trends in Pharmacological Science*, **15**, 71–6.

Gruchalla, R. S., Dinh, T. T. and Kennerly, D. A. (1990) An indirect pathway for receptor mediated 1,2-diacylglycerol formation in mast cells: IgE receptor mediated activation of phospholipase D. *Immunology*, **144**, 2334–42.

Hirata, M., Hayaishi, Y., Ushikubi, F. *et al.* (1991) Cloning and expression of cDNA for a human thromboxane A$_2$ receptor. *Nature*, **349**, 617–20.

Lin, L.-L., Wartmann, M., Lin, A. Y., Knopf, J. L., Seth, A. and Davis, R. J. (1993) cPLA$_2$ is phosphorylated and activated by cAMP kinase. *Cell*, **72**, 269–78.

Kumada, T., Miyata, H. and Nozawa, Y. (1993) Involvement of tyrosine phosphorylation in the IgE receptor mediated phospholipase D activation in rat basophilic
leukemia (RBL-2H3) cells. *Biochemical and Biophysical Research Communications*,
191, 1363–8.

McGiff, J. C. (1991) Cytochrome P-450 metabolism of arachidonic acid. *Annual
Review of Pharmacology and Toxicology*, **31**, 339–69.

Moolenaar, W. H., Jalink, K. and van Corven, E. J. (1992) Lysophosphatidic acid: a
bioactive phospholipid with growth factor-like properties. *Reviews of Physiology,
Biochemistry and Pharmacology*, **199**, 47–65.

Murakami, K. and Routtenberg, A. (1985) Direct activation of purified protein
kinase C by unsaturated fatty acids (oleate and arachidonate) in the absence of
phospholipids and Ca^{2+}. *FEBS Letters*, **192**, 189–93.

Natarajan, V., Taher, M. M., Roehm, B. *et al.* (1993) Activation of endothelial cell
phospholipase D by hydrogen peroxide and fatty acid hydroperoxide. *Journal of
Biological Chemistry*, **268**, 930–7.

Ordway, R. W., Singer, J. J. and Walsh, J. V. (1991) Direct regulation of ion channels by fatty acids. *Trends in Neurosciences*, **14**, 96–100.

Piomelli, D. (1993) Arachidonic acid in cell signalling. *Current Opinion in Cell Biology*, **5**, 274–80.

Piomelli, D. and Greengard, P. (1990) Lipoxygenase metabolites of arachidonic acid in
neuronal transmembrane signalling. *Trends in Pharmacological Sciences*, **11**, 367–73.

Qui, Z.-H., De Carvalho, M. S. and Leslie, C. C. (1993) Regulation of phospholipase A_2 activation by phosphorylation in mouse peritoneal macrophages. *Journal
of Biological Chemistry*, **268**, 24506–13.

Sharp, J. D., White, D. L., Chiou, X. G. *et al.* (1991) Molecular cloning and expression of human Ca^{2+} sensitive cytosolic phospholipase A_2. *Journal of Biological
Chemistry*, **266**, 14850–3.

Sigal, E. (1990) The molecular biology of mammalian arachidonic acid metabolism.
American Journal of Physiology, **260**, L13–L28.

Sugimoto, Y., Namba, T., Honda, A. *et al.* (1992) Cloning and expression of a
cDNA for mouse prostaglandin E receptor EP_3 subtype. *Journal of Biological
Chemistry*, **267**, 6463–6.

Sphingolipid Pathways

Ballou, L. R., Chao, C. P., Holness, M. A., Barker, S. C. and Raghow, R. (1992)
Interleukin 1 mediated PGE_2 production and sphingomyelin metabolism: evi-
dence for regulation of cyclo-oxygenase by sphingosine and ceramide. *Journal of
Biological Chemistry*, **267**, 20044–50.

Dobrowsky, R. T. and Hunnum, Y. A. (1992) Ceramide stimulates a cytosolic protein phospholipase. *Journal of Biological Chemistry*, **267**, 5048–51.

Hunnum, Y. A. (1994) The sphingomyelin cycle and the second messenger function
of ceramide. *Journal of Biological Chemistry*, **269**, 3125–8.

Hunnum, Y. A. and Bell, R. M. (1989) Functions of sphingolipids and sphingolipid
breakdown products in cellular regulation. *Science*, **243**, 500–7.

Hunnum, Y. A., Loomis, C. R., Merrill, A. H. and Bell, R. M. (1986) Sphingosine inhibition of protein kinase C activity and of phorbol dibutyrate binding *in vitro* and in human platelets. *Journal of Biological Chemistry*, **261**, 2604–9.

Merrill, A. H. and Jones, D. D. (1990) An update of the enzymology and regulation of sphingomyelin metabolism. *Biochimica et Biophysica Acta*, **1044**, 1-12.

Schütze, S., Potthoff, K., Machleidt, T., Berkovic, D., Wiegmann, K. and Krönke, M. (1992) TNF activated NF-kappa B by phosphatidylcholine specific phospholipase C induced acidic sphingomyelin breakdown. *Cell*, **71**, 765–76.

Spiegel, S. and Milstein, S. (1995) Sphingolipid metabolites: members of a new class of lipid second messengers. *Journal of Membrane Biology*, **146**, 225–37.

Related Lipid-derived Signalling Molecules

Anderson, J. M. (1991) Jasmonic acid dependent increase in vegetative storage protein in soybean tissue cultures. *Journal of Plant Growth Regulation*, **10**, 5–10.

Koda, Y. (1992) The role of jasmonic acid and related compounds in the regulation of plant development. *International Review of Cytology*, **135**, 155–99.

Koda, Y., Kikuta, Y., Tazaki, H., Tsujino, Y., Sakamura, S. and Yoshihara, T. (1991) Potato tuber inducing activities of jasmonic acid and related compounds. *Phytochemistry*, **30**, 1435–8.

Mueller-Uri, F., Parthier, B. and Nover, L. (1988) Jasmonate induced alteration of gene expression in barley leaf segments analysed by *in* vivo and *in vitro* protein synthesis. *Planta*, **176**, 241–7.

Weidhase, R. A., Lehmann, J., Kramell, H., Sembdner, G. and Parthier, B. (1987) Degradation of ribulose 1,5-bisphosphate carboxylase and chlorophyll in senescing barley leaf segments triggered by jasmonic acid methylester and counteraction by cytokinin. *Physiologia Plantarum*, **69**, 161–6.

Weidhase, R. A., Kramell, H., Lehmann, J., Liebisch, H., Lerbs, W. and Parthier, B. (1987) Methyljasmonate induced changes in the polypeptide pattern of senescing barley leaf segments. *Plant Science*, **51**, 177–86.

7 Intracellular Calcium: its Control and Role as an Intracellular Signal

Topics

- Calmodulin
- The Plasma Membrane and its Role in the Maintenance of Calcium Concentration
- Intracellular Stores
- Nerve Cells
- Gradients, Waves and Oscillations
- Sphingosine 1-Phosphate
- Cyclic ADP Ribose
- Nicotinate Adenine Dinucleotide Phosphate
- Apoptosis
- Fluorescence Detection and Confocal Microscopy

Introduction

Most intracellular signalling molecules are produced by the cell, have an effect and then are destroyed, for example the production of cAMP, its role in the activation of a kinase and its destruction by phosphodiesterase. However, this rationale is not always followed. Variation in the concentration of Ca^{2+} ions inside the cell, $[Ca^{2+}]_i$, is one of the major signals involved in the control of many metabolic processes. It is not its production or destruction that controls the signal, but the concentration experienced by the effector molecule in the particular compartment in which that effector acts that constitutes the signal. This signalling method is ubiquitous, being found in bacteria right through to mammals and is especially important in the neuronal pathways of higher animals.

The level of Ca^{2+} ions can be controlled in many ways, each of which has to work in a coordinated manner to allow the concentration of Ca^{2+} to be an effective signal. The steady-state level of Ca^{2+} in the cytoplasm of a cell is usually less than 10^{-7} mol/l. If the cell is activated by an influx of Ca^{2+}, this level might rise to as high as 10^{-5} mol/l. To maintain this very low steady-state level, with an extracellular Ca^{2+} concentration that may well be 2×10^{-3} mol/l, Ca^{2+} ions are actively pumped out of the cell by enzymes on the plasma membrane. Cells also sequester Ca^{2+} in the endoplasmic reticulum (ER), or sarcoplasmic reticulum in muscle tissue, or in the mitochondria. Again this is an active process and therefore a considerable amount of the cell's energy is devoted to the control of the intracellular Ca^{2+} concentration. A signal is propagated when this Ca^{2+} is released back into the cytoplasm, so rapidly increasing the Ca^{2+} concentration in the desired area. This release is often under the control of a second signalling molecule, for example $InsP_3$, which is itself released from the plasma membrane by the breakdown of $PtdInsP_2$, as discussed more fully in Chapter 6. $InsP_3$ binds to $InsP_3$ receptors on the membranes of the intracellular stores. These receptors are also Ca^{2+} channels that allow the rapid release of Ca^{2+} back into the cytoplasm of the cell. An example of such a signal-transduction pathway is shown in Figure 7.1.

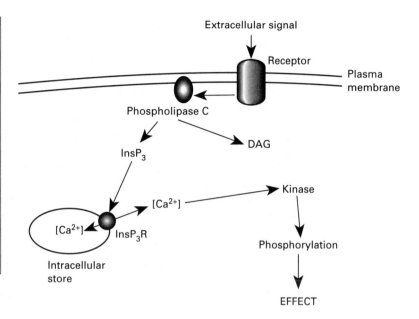

Figure 7.1
An example of how ligand binding leads to release of Ca^{2+} from intracellular stores. Activation of phospholipase C gives rise to the messenger inositol trisphosphate ($InsP_3$), which binds to $InsP_3$ receptors ($InsP_3R$) of the endoplasmic reticulum membranes. Such receptors are also channels and open to release Ca^{2+} ions. DAG, diacylglycerol.

The third way in which Ca^{2+} concentrations are controlled in the cytoplasm is by the presence of Ca^{2+}-binding proteins. Often the action of Ca^{2+} on the effector molecule, such as kinases, is through such a binding protein. The ubiquitous example of such a protein is calmodulin. Ca^{2+} ions can coordinate up to 12 oxygen atoms, but coordination numbers commonly range between six and eight. This is exploited by many binding proteins, which bind Ca^{2+} ions through six or seven oxygen atoms, usually provided by glutamate or aspartate residues. These amino acids are normally charged at physiological pH and therefore able to partake in this role. A second common feature of many Ca^{2+}-binding proteins is the presence of a structure known as an EF hand. This structure gained its name from the E and F regions of the protein parvalbumin, which serves as a Ca^{2+} buffer in muscle. Parvalbumin has an amino acid secondary structure that comprises a series of α-helices, labelled by letter. The α-helices E and F are positioned in such a way as to point like the forefinger and thumb of a right hand, hence the name EF hand. A Ca^{2+}-binding loop containing active Ca^{2+}-binding glutamate and aspartate residues lies between these α-helices. Many Ca^{2+}-binding proteins contain more than one binding site, and it is not unusual to see cooperative binding.

These binding proteins can potentially fill two roles in the formation of a Ca^{2+} signal. They might simply be acting as buffers, so that when the Ca^{2+} concentration alters free Ca^{2+} ions are not left in solution to have an effect. Alternatively, the proteins might act as the trigger, mediating the rise in free Ca^{2+} and turning it into the signal. Many of these proteins undergo a major alteration in their conformation when they bind to Ca^{2+}, and this conformational change undoubtedly exposes active sites on the protein and so allows further reactions and propagation of the signal.

145

One of the most well-studied and best-characterised examples of a Ca^{2+}-binding protein with a clearly defined role is troponin. Troponin is, in fact, a complex of three polypeptides, TnC, TnI and TnT, of 18, 24 and 37 kDa respectively. Along with the protein tropomyosin, troponin lies along the thin filaments of muscle fibres. The TnC subunit binds Ca^{2+} ions. It has two homologous domains, and each has two EF hand-binding sites for Ca^{2+}, those in the C-terminal domain being of high affinity while those in the N-terminal domain are of lower affinity. The domains are held together by an α-helix, the whole molecule showing great structural similarity to calmodulin. On binding of Ca^{2+} to the low-affinity sites, i.e. when Ca^{2+} has been released from the sarcoplasmic reticulum, a conformational change takes place in the TnC polypeptide, which is transmitted through the other polypeptides in the complex and to tropomyosin. A change in the orientation of tropomyosin relieves the steric hindrance it causes to actin–myosin interaction, allowing muscle function.

Other Ca^{2+} binding proteins include a protein called calbindin-D_{28k}. This is a 28k-Da polypeptide that has a wide but heterogeneous distribution in the brain. Another important example is calsequestrin. This 44-kDa protein is surprisingly acidic with over one-third of its amino acids as either aspartic acid or glutamic acid, and amazingly, each calsequestrin binds 43 Ca^{2+} ions. It is found in the sarcoplasmic reticulum where it controls the concentration of free Ca^{2+} and so helps to reduce the Ca^{2+} gradient across the membrane. Also present here is a polypeptide known as high-affinity Ca^{2+}-binding protein.

But what is Ca^{2+} controlling in the cell? To answer that question would take a long time as Ca^{2+} seems to have a plethora of actions. Many of the actions of Ca^{2+} ions have been elucidated by the use of calcium ionophores, such as A23187 or ionomycin. Such chemicals effectively punch holes in the plasma membrane and let Ca^{2+} ions flood into the cell, so mimicking the sudden rise of Ca^{2+} concentration that the cell achieves by the release of intracellular stores or via a signal-mediated response. Alternatively, Ca^{2+} ions may be removed from the extracellular medium by the addition of chelating agents, such as EDTA, which prevent the cell from using plasma-membrane channels as a means to increase cytoplasmic Ca^{2+}. By the use of such methods the diversity of the roles of Ca^{2+} ions in the signalling of cells can be discerned. Some of the proteins known to have their activity modulated by Ca^{2+}-signalling systems are listed in Table 7.1. Ca^{2+} is certainly involved in the activation of Ca^{2+}/calmodulin-

Table 7.1. Some of the proteins controlled by Ca^{2+} ion concentrations.

Ca^{2+}-dependent protein kinase
Ca^{2+}-dependent protein phosphatase (for example, PP2B)
Phospholipase C (particularly δ form)
Pyruvate dehydrogenase
Nitric Oxide synthase
Protein kinase C
Adenylyl cyclase (particularly type 1)
Inositol trisphosphate receptors and ryanodine receptors
Ca^{2+}/Mg^{2+}-dependent endonuclease
Phosphorylase kinase
Ca^{2+}-ATPase
Troponin

dependent protein kinase, PKC, Ca^{2+}-dependent phosphatases and in the activation of nitric oxide synthase leading to the propagation of the intracellular signal. It is also involved in the direct activation of many enzymes, such as pyruvate dehydrogenase, as well as in more complex patterns of activity such as the control of the cell cycle. Spikes of Ca^{2+} have been shown to be essential for triggering the completion of meiosis and the initiation of mitosis in some cells, while depletion of Ca^{2+} stores can arrest cells in the G_0–G_1 phases. Gene expression can be under the control of Ca^{2+} concentrations, and here again Ca^{2+}-binding proteins such as calreticulin may mediate the changes in concentration into the final signal.

Calmodulin

Calmodulin is involved in the mediation of the Ca^{2+} signal in a vast number of regulatory pathways, including the regulation of metabolic activity, the regulation of other signal pathways, such as NO generation, and the regulation of gene expression. It is sometimes found as an integral part of an enzyme complex, as in phosphorylase kinase, where it is known as the δ subunit. Often, it even appears to be involved in the contradictory regulation of an activity, responsible for both an increase and a decrease in the signal at the same time. One example of this can be seen with its role in the control of cAMP levels, where it can increase the production of cAMP via the activation of adenylyl cyclase while at the same time activating its breakdown by the stimulation of Ca^{2+}/calmodulin-dependent phosphodiesterase. Calmodulin can also lead to the activation of plasma-membrane and ER Ca^{2+} pumps that causes the cessation of the Ca^{2+} signal.

Calmodulin is a small, relatively acidic protein that has the capacity to bind four molecules of Ca^{2+}. It has an affinity for Ca^{2+} of approximately 10^{-6}mol/l, which lies between the resting Ca^{2+} concentration of the cell and the activated concentration, which might reach levels as high as 10^{-5}mol/l. The shape of the protein molecule resembles a dumb-bell, with two globular regions connected by a long, flexible and very mobile α-helix (Figure 7.2). Each of the globular domains contains two EF hand regions, with their characteristic helix–loop–helix topology, each of which can bind to one molecule of Ca^{2+}. The two EF hands within each globular region are connected by a short antiparallel β-sheet region. Interestingly, despite the similarity in the two regions, their affinity for Ca^{2+} differs. The C-terminal domain has the higher affinity and binds to Ca^{2+} first, leading to major conformational changes within the molecule. These structural changes reveal two hydrophobic patches, one in each half of the protein. It is these patches that interact with the molecule that calmodulin controls.

With the use of synthetic peptides to block binding, proteolysis experiments to remove potential binding sites from proteins and the expression of cDNAs, it has been attempted to map the areas on proteins that bind to the hydrophobic patches revealed in calmodulin after it has bound to Ca^{2+}. Most of these sites involve an α-helix of 16–35 amino acids. If the α-helix is viewed from one end, it can been seen that it possesses polarity, with the majority of the polar, mainly basic, amino acids

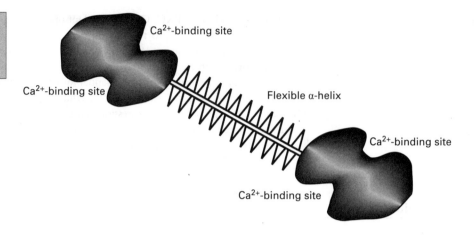

Figure 7.2
Domain structure of calmodulin showing the Ca^{2+}-binding sites.

lying down one side and the majority of the hydrophobic amino acids lying on the opposite side (Figure 7.3) The binding regions also usually contain a potential phosphorylation site.

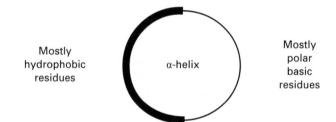

Figure 7.3
End-on view of an α-helix that may be involved in the binding of a protein to activated calmodulin.

Besides the regulation of calmodulin by the binding of Ca^{2+} to the four binding sites, calmodulin is itself phosphorylated *in vivo*. The central helix is phosphorylated at two sites, with a third site located in the third Ca^{2+}-binding region. It is thought that the interaction of calmodulin with its target protein is reduced on phosphorylation and this might be caused by inhibition of the structural changes that occur in the calmodulin molecule on Ca^{2+}-binding, with the hydrophobic patches not being revealed properly.

The Plasma Membrane and its Role in the Maintenance of Calcium Concentration

The plasma membrane contains pumps that actively discharge Ca^{2+} from the cell, helping to maintain the low cytoplasmic Ca^{2+} concentration. The plasma membrane Ca^{2+}-ATPase (PMCA) pump is a P-type ATPase, using the energy of one or two molecules of ATP to drive the transport of one molecule of Ca^{2+} out of the cell. The pumping cycle uses a phosphorylated intermediate, involving the transient transfer of a phosphoryl group to an aspartate residue on the polypeptide. This phosphorylation of the protein causes the conformational change needed to transport Ca^{2+} ions across the membrane; dephosphorylation restores the protein to its former state ready for another round of Ca^{2+} pumping. The net result is the breaking and loss of the high-energy phosphate bond of ATP, the energy being used to drive the Ca^{2+}-pumping activity of the ATPase.

The ATPase protein probably has 10 transmembrane-spanning domains, with tissue-specific isoforms arising from four genes as well as alternative splicing of gene transcripts. Control of the PMCA pump may be through a variety of factors. Calmodulin causes a feedforward type of activation of the pumps, Ca^{2+} ion activation of calmodulin leading to both an increase in the V_{max} of the pump as well as a lowering of the K_m of the pump for Ca^{2+}, resulting in greater extrusion of Ca^{2+} by the pumps if cytosolic Ca^{2+} rises. Control is also seen through the action of kinases, such as cAPK or PKC, while the activity of PMCAs is also influenced by the presence of phospholipids.

However, this is not the only pumping system involved. The Na^+/Ca^{2+} exchanger is also active in the expulsion of Ca^{2+} ions from the cell (Figure 7.4).

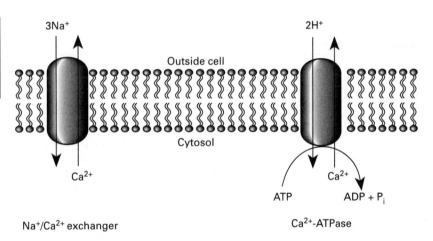

Figure 7.4
Plasma-membrane Ca^{2+} pumps: an Na^+/Ca^{2+} exchanger and a Ca^{2+}-ATPase.

3Na$^+$

2H$^+$

Outside cell

Cytosol

Ca^{2+}

Ca^{2+}

ATP

ADP + P$_i$

Na$^+$/Ca^{2+} exchanger

Ca^{2+}-ATPase

The Ca^{2+}-ATPase has a higher affinity for Ca^{2+} ions than the Na^+/Ca^{2+} exchanger, but the latter can pump Ca^{2+} ions at a much higher rate. A typical Ca^{2+}-ATPase has been shown to pump Ca^{2+} ions at a rate of approximately 30/s in an *in vitro* situation, whereas the Na^+/Ca^{2+} exchanger can pump Ca^{2+} ions at a rate up to 2000/s. Therefore, large amounts of Ca^{2+} can be expelled from the cell quite quickly. When the Ca^{2+} ion concentration is in the micromolar range, the slower but higher affinity Ca^{2+}-ATPase continues to decrease the cytoplasmic Ca^{2+} ion concentration further.

The plasma membrane is also a major site of re-entry of Ca^{2+} back into the cell. In most cells re-entry is controlled by the polarisation of the membrane as well as the presence of other second-messenger molecules. The opening of potassium (K^+), channels alters the membrane potential across the plasma membrane and this drives

Ca^{2+} ions into the cytoplasm through voltage-independent Ca^{2+} channels. The structure of the channels is such that re-entry is very selective for Ca^{2+} ions, not allowing the entry of other ions.

The depletion of intracellular stores can also serve as a signal for the re-rentry of Ca^{2+} through the plasma membrane. Work has intensified to find the channels involved. One that has been extensively studied is responsible for the Ca^{2+}-release-activated current (I_{CRAC}) or depletion-activated current (I_{DAC}). Recently, a cDNA has been isolated from bovine tissues that codes for a protein designated as CCE (capacitative Ca^{2+} entry) and which shows significant homology to the transient receptor potential (*trp*) gene product from *Drosophila*. This protein is thought to be involved in Ca^{2+} re-entry in mammalian tissues.

Many second-messenger molecules have been suggested as the signal for the opening of plasma-membrane channels, including $InsP_3$, tyrosine phosphorylation, cGMP and G proteins, both the trimeric and small monomeric ones, but the exact mechanism has yet to be discovered. A new messenger, named Ca^{2+} influx factor (CIF), has been suggested as a candidate for this role. It has been characterised as a phosphorylated anion of less than 500 Da, but more studies will be required to fully elucidate its mode of action.

Intracellular Stores

Endoplasmic Reticulum Stores

A crucial aspect of a Ca^{2+} ion signalling is the maintenance of a very low cytoplasmic Ca^{2+} concentration and the ability to allow it to suddenly rise. Both these aspects of the signalling process are facilitated by the presence of intracellular stores of Ca^{2+}, in which Ca^{2+} can be sequestered and from which Ca^{2+} can be rapidly released

One of the major stores is the lumen of the ER, or in muscle tissues the sarcoplasmic reticulum. Specialised areas of ER used for Ca^{2+} storage have been observed in some cells and have been called calciosomes.

In the membrane of the ER are Ca^{2+} pumps that pump Ca^{2+} from the cytoplasm to the inside of the membranous network (Figure 7.5) These are commonly referred to as smooth endoplasmic reticulum Ca^{2+}-ATPase (SERCA) pumps. In the specialised type of ER found in muscle, the sarcoplasmic reticulum, they are in fact a major protein component of the membrane, being approximately 80% of the integral proteins found in that membrane. The density of these pumps has been estimated to be as high as $25\,000/\mu m^2$. As the name suggests, this is an active process, pumping Ca^{2+} from a region of very low concentration to a region of relatively high concentration, using the breakdown of ATP as an energy source. Again, like the plasma-membrane equivalent, a phosphoryl intermediate is involved in the pumping steps, i.e. a phosphate group is added to an aspartate residue on the polypeptide causing a conformational change, resulting in Ca^{2+} ions moving across the membrane where they can be released. Removal of the phosphate group restores the enzyme to its first conformation. It is thought that two Ca^{2+} ions are transported for every ATP hydrolysed.

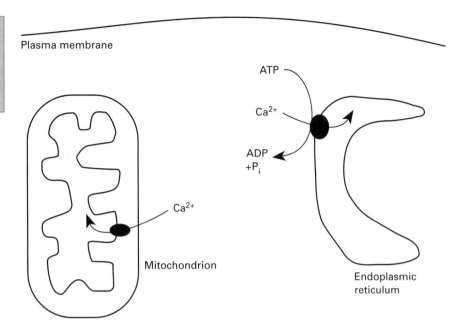

Figure 7.5
Intracellular Ca^{2+} stores:
mitochondria and the
endoplasmic reticulum
both sequester Ca^{2+}
ions.

Plasma membrane

ATP

Ca^{2+}

ADP
+P$_i$

Ca^{2+}

Mitochondrion

Endoplasmic
reticulum

Therefore, like the PMCA (pumping one Ca^{2+}/ATP), the SERCA (pumping two Ca^{2+}/ATP) is classified as a P-type ATPase with 10 transmembrane-spanning regions defined in the polypeptide. However, despite this topological and functional similarity, very little sequence homology exists between the two types of Ca^{2+} pumps. The endoplasmic protein appears to contain two main domains, with a smaller one on the luminal side of the membrane. One main domain is buried in the membrane and acts as a Ca^{2+} channel, the Ca^{2+} ions binding near the centre line of the membrane. The other major domain contains the ATPase activity on the cytoplasmic side of the membrane and the site of phosphorylation.

When a functional Ca^{2+}-ATPase is reconstituted into a phospholipid membrane, it is seen to pump Ca^{2+} ions at a rate of approximately 30/s; interestingly, if a Ca^{2+} gradient is artificially set up across the membrane the ATPase can be forced to run in reverse and ATP formation is seen if ADP is present.

SERCA pumps arise from three genes that appear to be expressed in a tissue-specific manner. The gene products SERCA1 and SERCA2 are expressed in different types of muscle, while SERCA3 is expressed in non-muscle tissue. Thapsigargin, a lactone that contains tumour-promoting activity, has been identified as a highly specific inhibitor of SERCA activity, trapping the pump in a Ca^{2+}-free state and has been useful as a biochemical tool for the study of these activities as it does not inhibit plasma-membrane pumps. Alternatively, orthovanadate can be used as a more general inhibitor of P-type ATPases.

The activity of Ca^{2+}-ATPases is probably not continuous but is controlled. An increase in cytoplasmic Ca^{2+} accelerates the activity of the ER pump, while an increase of free Ca^{2+} inside the ER inhibits the activity of the pump, half the activity being lost if the Ca^{2+} concentration reaches approximately 300 µmol/l. Control of the pumps may also come from a receptor-mediated mechanism, but the details await elucidation.

151

Once inside the ER the Ca^{2+} is sequestered and buffered, keeping the concentration of free ions down. One such sequestering protein is calsequestrin. No role for the sequestering of Ca^{2+} ions has been assigned other than as a store that can be released back to the cytoplasm at the appropriate moment. However, as one of the major roles of the ER is in the sorting and trafficking of proteins, it has been suggested that Ca^{2+} ions might have a dual role here, although there is little evidence of this.

As discussed in the last chapter, one of the products of $PtdInsP_2$ breakdown is $InsP_3$, the major role of which is the release of Ca^{2+} from intracellular stores. The ER contains $InsP_3$ receptors that trigger the release of the Ca^{2+}. The receptors are tetrameric in conformation with a C-terminal domain spanning the membrane of the ER. Each subunit has an approximate molecular mass of 310 kDa and contains a cationic pore, which shows relatively little selectivity. Each of the four subunits contains an arginine- and lysine-rich region at the N-terminal end that binds one $InsP_3$ molecule. The N-terminal region, which is on the cytoplasmic side of the membrane, also has binding sites for two ATP molecules and at least one Ca^{2+} ion. As is regularly seen with other signalling proteins, the $InsP_3$ receptor can exist in several isoforms. At least four genes encode $InsP_3$ receptors with significant homology between them. Expression of the receptors, as might be expected, differs between different tissues.

The receptors' activity is not only controlled by the binding of $InsP_3$ but very importantly the release of Ca^{2+} through these receptors is modulated by Ca^{2+} itself, Ca^{2+} being referred to as a co-agonist. If low concentrations of $InsP_3$ are present then at low cytoplasmic Ca^{2+} concentrations Ca^{2+} release is accelerated but at high cytoplasmic Ca^{2+} concentrations the release of Ca^{2+} through $InsP_3$ receptors is inhibited. Therefore, if plotted, the release response to Ca^{2+} concentrations would give a bell-shaped curve, with a maximum around 0.2–0.3 µmol/l (Figure 7.6).

Figure 7.6
Ca^{2+}-induced Ca^{2+} release from inositol trisphosphate ($InsP_3$) receptors in the presence of lower concentrations of $InsP_3$.

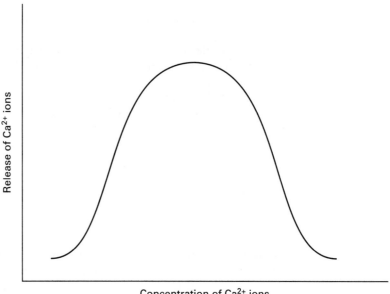

Release of Ca^{2+} ions

Concentration of Ca^{2+} ions

Control of InsP$_3$ receptors may also be via phosphorylation through cAPK and its activity is also modulated by the presence of Mg^{2+}. Caffeine has been shown to inhibit Ca^{2+} release from InsP$_3$ receptors in response to InsP$_3$, while heparin competitively inhibits InsP$_3$ receptors by binding to the InsP$_3$ binding sites. Interestingly, ethanol is also a potent inhibitor of InsP$_3$ receptors.

It is also interesting to note that InsP$_3$ receptors have also been found in the plasma membrane of some cells, but their role in this location has not been elucidated.

As well as InsP$_3$, other inositol phosphate metabolites might be involved in the release and control of Ca^{2+}. For example, InsP$_4$ has been shown to affect Ca^{2+} influx in cells and also to be an inhibitor of Ca^{2+}-ATPase pumps.

The other Ca^{2+} ion release channel found in the ER is the ryanodine receptor (RyR). Like the InsP$_3$ receptor, the RyR is also a tetramer but the subunits are much bigger, having a molecular mass of approximately 560 kDa, and the receptor has been purified as a 30S complex. Three genes (*ryr-1*, *ryr-2* and *ryr-3*) code for separate isoforms of the RyR, with expression of the three being very tissue specific. The gene *ryr-1* is expressed in skeletal muscle, while *ryr-2* is found expressed in cardiac muscle. The gene *ryr-3* represents a non-muscle form. Of the proteins encoded by these genes, RyR1 and RyR2 show a large amount of sequence homology but RyR3 is much smaller, at less than 650 amino acids. The C-terminal ends of the subunits create the Ca^{2+} channel through the membrane, having between four and ten putative membrane-spanning regions, while the N-terminal end protrudes into the cytosol. As the name suggests, these channels are sensitive to ryanodine, a plant alkaloid. Low concentrations of ryanodine cause opening of the RyR channels but at higher, micromolar, concentrations of ryanodine the channels are once again closed. Opening of the RyR is also seen in the presence of caffeine.

What causes the RyR to open as a physiological response? In some tissues, for example striated muscle, the sarcoplasmic reticulum comes into close contact with the plasma membrane and protein complexes exist that connect the two membranes. These protein complexes are known as T-tubule feet and contain RyRs. The other part of the foot structure is a plasma membrane-associated voltage-sensitive receptor, such as the dihydropyridine receptor. This receptor senses changes in the voltage across the membrane and undergoes a conformational change. This change in conformation of the plasma membrane-associated receptor is sensed by RyR, as it is part of the same protein complex. A conformational change is induced in the RyR that results in channel opening and the release of Ca^{2+} ions (Figure 7.7a).

Alternatively, like InsP$_3$ receptors, the RyR is also sensitive to low concentrations of Ca^{2+}, having a similar bell-shaped dependence on Ca^{2+}. Like InsP$_3$ receptors, low concentrations of Ca^{2+} induce the opening of the RyR while higher, millimolar, concentrations inhibit. Therefore, a voltage change across the plasma membrane may induce a small influx of Ca^{2+} from the outside, through a Ca^{2+} voltage-gated channel. This is sufficient to induce the opening of the RyR and the release of intracellular Ca^{2+} stores (Figure 7.7b).

Figure 7.7
Mechanisms for release of Ca^{2+} through ryanodine receptors (RyR). (a) Voltage changes are sensed by a plasma membrane-associated receptor and protein conformational changes are relayed to the ryanodine receptor, causing is opening. (b) Voltage-gated Ca^{2+} channels allow small amounts of Ca^{2+} to enter the cell, which trigger ryanodine receptors to open.

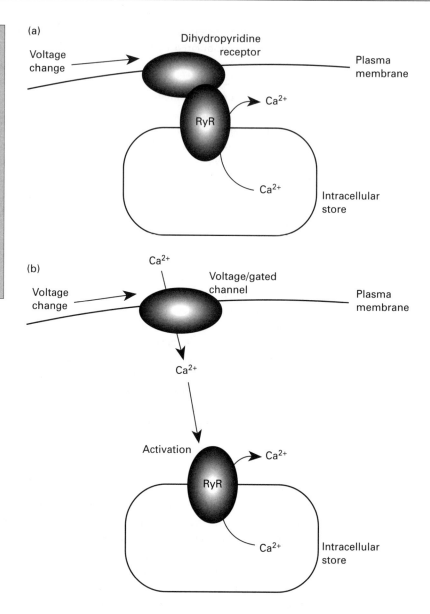

There are reports that the RyR is sensitive to, and opens in, the presence of cyclic ADP ribose (cADPR). RyR2 seems to be the target for cADPR as RyR1 is insensitive to this messenger. It is possible that RyR3 is also controlled in this way. The production and role of this putative messsenger is discussed further below. Control of the RyR by calmodulin and phosphorylation has also been reported, although the exact influence of these has yet to be established.

Interestingly, all the Ca^{2+} channels so far studied contain the amino acid sequence Thr-X-Cys-Phe-Ile-Cys-Gly. This is found in the C-terminal ends of the polypeptides and may play a role in the opening of the channel. It also seems to bestow thiol sensitivity to the channels.

Mitochondrial Calcium Metabolism

As well as sequestration of Ca^{2+} by the ER, other organelles also actively take up Ca^{2+} ions. One of the most significant are the mitochondria, which acquire Ca^{2+} through the use of a Ca^{2+} uniporter, driven by the proton gradient across the inner mitochondrial membrane that is maintained by the respiratory electron transport chain (Figure 7.5). Levels inside the mitochondria can rise to approximately $0.5\,\text{mM}$ but the affinity of the pump for Ca^{2+} is less than that of the ER.

Ca^{2+} in the mitochondria is certainly used to control the activity of enzymes, such as pyruvate dehydrogenase, which acts as a crucial link between glycolysis and the citric acid cycle and hence is involved in the production of ATP. Ca^{2+} ion concentrations in the mitochondria have also been linked to the onset of degenerative diseases, where a lack of control can lead to a depletion of ATP in neuronal tissue leading to neuronal death.

Nerve Cells

The plasma membrane re-entry of Ca^{2+} ions in neuronal tissue differs from that of other cells. Nerve cells contain voltage-gated Ca^{2+} ion channels. Depolarisation of the resting membrane potential causes a conformational change in Ca^{2+}-selective ion channels. The proteins contain special regions, S4 regions, which act as voltage-sensing regions. The membrane potential then supplies the driving force for the movement of Ca^{2+} ions through the open channels. To a certain extent the process is self-controlling, as the channels appear to close in a time-dependent manner; if the membrane is once again depolarised then the driving force, the membrane potential, is depressed and does not supply the energy for Ca^{2+} movement.

The internal stores of Ca^{2+} in neurones are released mainly through the RyR, as discussed above for Ca^{2+} release from the ER.

Gradients, Waves and Oscillations

An interesting and challenging aspect of the Ca^{2+} signalling story is the fact that the concentration of Ca^{2+} is not uniform throughout the cytoplasm, or indeed in the compartment in which it might be sequestered. Intuitively, it would be expected that if Ca^{2+} ions rush in from the outside of the cell through the plasma membrane then the concentration would rapidly dissipate throughout the cytoplasm by diffusion. However, this is clearly not the case. It has been estimated that Ca^{2+} ions only migrate for approximately $50\,\mu s$, a distance of only $0.1\text{-}0.5\,\mu m$. In that time the ions are bound by local binding proteins. However, such estimations are altered by the saturation of the binding proteins and by the uneven distribution of such proteins. Some buffering proteins might be free in the cytoplasm while others might be bound and immobile, again altering the equation. Therefore a gradient of Ca^{2+} concentration may extend away from the site of Ca^{2+} release. Such gradients are called micro-gradients as they occur within a space of 1-$10\,\mu m$. The highest concentrations of Ca^{2+} may reach up to $1\,\mu mol/l$. As well as these micro-gradients, very high local concentrations are found around an open channel. Mobilisation through a channel may well be more efficient than the diffusion of Ca^{2+} ions through the cytosol, resulting in a very steep gradient extending away from that channel. These gradients decay within tens of nanometres of the channel opening and are therefore referred to as nano-gradients. Typically such nano-gradients only last for less than a millisecond but could have Ca^{2+} concentrations as high as $100\,\mu mol/l$. Gradients of Ca^{2+} have been proposed to be important in the control of cell migration, exocytosis, ion transport and gap-junction regulation.

One phenomenon that remains a puzzle is the observation that Ca^{2+} concentrations can be measured as oscillations or waves. Therefore, it might not be the overall concentration of Ca^{2+} that actually constitutes the signal but rather the amplitude and frequency of the concentration wave or oscillation.

A typical mechanism for the creation of a wave is described below. It is probable that local concentrations of Ca^{2+} are created around the opening of channel in the ER which in turn may cause the activation of channels in the locality, so releasing Ca^{2+} further along the membrane. Simultaneously, such local concentrations inhibit the first channel from releasing any more Ca^{2+} while the local Ca^{2+}-ATPases start to actively remove the ions, once again lowering the Ca^{2+} concentration in the location of the first channel to open. The new release further along the membrane can then initiate the same response with the channels and ATPases even further along the membrane, and so the wave propagates out from the site of initiation. Such a scheme could propagate a wave of Ca^{2+} across the membrane (Figure 7.8). Interestingly, it has been reported that over expression of the Ca^{2+}-ATPase pump of the ER increases the frequency of the concentration waves recorded, while waves have been reported to travel from cell to cell through gap junctions; this may constitute a longer range signalling mechanism than first envisaged. Such waves may move through the cytosol at speeds of 5–$100\,\mu m/s$ and, interestingly, wave speed may depend on the concentration of agonist applied to the cell.

As well as waves and gradients, Ca^{2+} concentrations in cells have been seen to oscillate. This was first demonstrated in blowfly salivary glands where it was observed that the Ca^{2+} concentration could be measured as a series of spikes and that the frequency of the spikes increased as the agonist concentration increased. Subsequent research has revealed that the patterns of the oscillation in fact depend on many factors, including the agonist concentration, the receptor type activated and the Ca^{2+}-buffering capacity of the cell.

How do such oscillations occur? Several models have been put forward to try to explain this phenomenon. One such model is known as the $InsP_3$–Ca^{2+} cross-coupling model (ICC model). An extracellular signal leads to the production of $InsP_3$ (see Figure 7.1) and this leads to the subsequent rise in cytoplasmic Ca^{2+} concentration. The increase in cytoplasmic Ca^{2+} ions then stimulates the activity of PLC leading to a greater release of Ca^{2+} from intracellular stores. However, the $InsP_3$ receptor is inhibited by high concentrations of Ca^{2+}, and therefore, once the cytoplasmic Ca^{2+} reaches a critical level, the $InsP_3$ receptor channels are closed and no further Ca^{2+} release takes place. The cell now actively removes the Ca^{2+} ions from the cytoplasm and so restores the cell to the unstimulated situation. The reduction in cytoplasmic Ca^{2+} also leads to a

Figure 7.8
Proposed scheme to explain the propagation of Ca^{2+} ion waves along the membranes of cells.

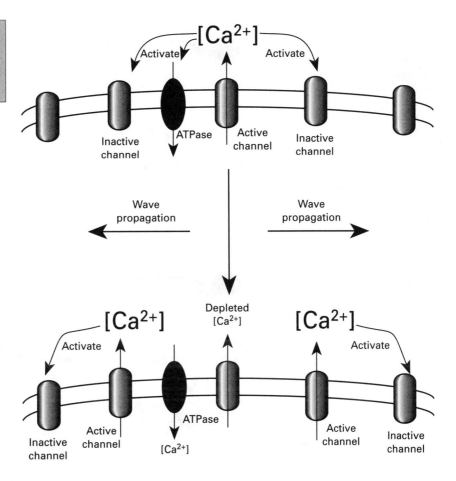

reduction of PLC activity and a reduction of $InsP_3$ in the cytoplasm. The presence of agonist can once again initiate a spike in Ca^{2+} release. In this model, the release of Ca^{2+} from intracellular stores is a consequence of the spiking in PLC activity. However, research has shown that even if the cytoplasmic $InsP_3$ concentration is kept constant the Ca^{2+} ion concentration still oscillates. Therefore further models are needed.

Two further models are based on the proposed presence of a single pool of releasable Ca^{2+}, the one-pool model, or the presence of two independent pools of releasable Ca^{2+}, the two-pool model. In the one-pool model, the initial release of Ca^{2+} is relatively slow. However, because opening of $InsP_3$ receptors by $InsP_3$ is enhanced by the presence of Ca^{2+}, as the cytoplasmic Ca^{2+} rises so the release of more Ca^{2+} through $InsP_3$ receptors is accelerated. This gives the rapid rise of Ca^{2+} seen as the early part of the spike. However, higher concentrations of Ca^{2+} inhibit $InsP_3$ receptors and the channels close, allowing no further rise in cytoplasmic Ca^{2+}. A drop is seen as the Ca^{2+} is resequestered and removed from the cytoplasm by the various Ca^{2+} pumps (Figure 7.9). It has also been suggested that the rise in Ca^{2+} causes the activation of PLA_2, with the concomitant release of AA. This could also potentially inhibit $InsP_3$ receptors and so play a role in the formation of the Ca^{2+} spike.

157

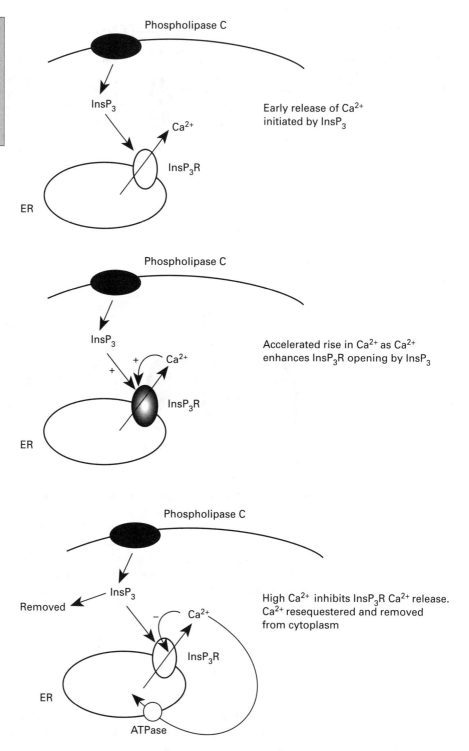

Figure 7.9
Schematic representation of the one-pool model of Ca^{2+} oscillations. ER, endoplasmic reticulum; InsP$_3$, inositol trisphosphate; InsP$_3$R, InsP$_3$ receptors.

Phospholipase C

InsP$_3$

Ca^{2+}

InsP$_3$R

ER

Early release of Ca^{2+} initiated by InsP$_3$

Phospholipase C

InsP$_3$

Ca^{2+}

InsP$_3$R

ER

Accelerated rise in Ca^{2+} as Ca^{2+} enhances InsP$_3$R opening by InsP$_3$

Phospholipase C

InsP$_3$

Removed

Ca^{2+}

InsP$_3$R

ER

ATPase

High Ca^{2+} inhibits InsP$_3$R Ca^{2+} release. Ca^{2+} resequestered and removed from cytoplasm

The two-pool model suggests that Ca^{2+} may be sequestered into different intracellular stores, which have receptors that differ in their sensitivity to $InsP_3$. In this model, only the most sensitive $InsP_3$ receptors cause a release from those independent stores (Figure 7.10). Other stores not releasing Ca^{2+} still actively accumulate Ca^{2+} and so their internal Ca^{2+} concentrations rise, while cytoplasmic Ca^{2+} buffers also become saturated. Once the concentration inside the second stores reaches a critical level, their membrane pumps are shut off and the cytoplasmic Ca^{2+} level rises, as no more Ca^{2+} can be buffered and no more can be sequestered. Above a threshold cytoplasmic Ca^{2+} concentration, all the $InsP_3$ receptors open and all the stores release Ca^{2+}. Hence the cytoplasmic Ca^{2+} ion concentration rises sharply, seen as the early phase of the spike. Again, as in the one-pool model, a high cytoplasmic Ca^{2+} concentration results in the inhibition of $InsP_3$ receptors and the cessation of Ca^{2+} release. Pumps in the membranes now actively resequester and remove Ca^{2+} from the cytoplasm, seen as the downslope of the spike.

A modification of the two-pool model postulates that one store contains $InsP_3$ receptors while the other contains RyRs. It must also be remembered that it is not only the release and pumping of Ca^{2+} at the intracellular store membranes that is important, but the plasma membrane plays a vital role in these events too.

Sphingosine 1-phosphate

As well as the classical route for Ca^{2+} release through the production of $InsP_3$, another lipid metabolic pathway also leads to the release of Ca^{2+} from intracellular stores, the sphingosine pathway (see Chapter 6). In this pathway, one of the main active second messengers appears to be sphingosine 1-phosphate, which can mobilise Ca^{2+} from internal cell stores in an inositol-independent manner and in a way independent of extracellular Ca^{2+}. Sphingosine 1-phosphate is produced from sphingosine by sphingosine kinase, an enzyme which, as expected, is dependent on ATP as a source of phosphoryl groups and which can be competitively inhibited by DL-threo-dihydrosphingosine. It is likely that the kinase is located within the ER membrane, where sphingosine 1-phosphate has its action.

Another metabolite reported to be important in Ca^{2+} ion oscillations in the cell is sphingosylphosphorylcholine (SPC). Although it seems to be the same Ca^{2+} stores released by this metabolite as those released by $InsP_3$, the response is not mediated by $InsP_3$ receptors.

Cyclic ADP Ribose

cADPR is synthesised by the enzyme ADP-ribosyl cyclase, which uses oxidised nicotinamide adenine dinucleotide (NAD^+) as a substrate. This enzyme is found in both invertebrate and mammalian cells. cADPR releases Ca^{2+} in sea-urchin eggs and may be an agonist for RyR2 and maybe RyR3, and therefore it has been pro-

Figure 7.10
Two-pool model of Ca^{2+} oscillations. Here, the two pools have inositol trisphosphate $(InsP_3)$ receptors $(InsP_3R)$ of differing sensitivities, but other models using two Ca^{2+} pools have been proposed.

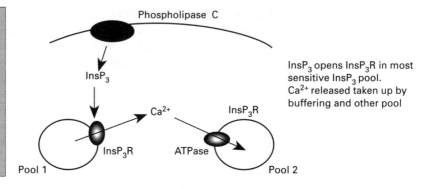

$InsP_3$ opens $InsP_3R$ in most sensitive $InsP_3$ pool. Ca^{2+} released taken up by buffering and other pool

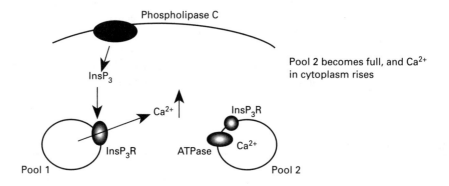

Pool 2 becomes full, and Ca^{2+} in cytoplasm rises

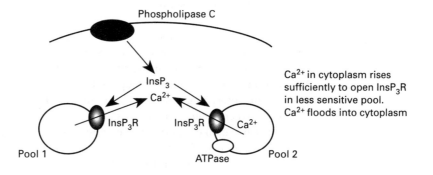

Ca^{2+} in cytoplasm rises sufficiently to open $InsP_3R$ in less sensitive pool. Ca^{2+} floods into cytoplasm

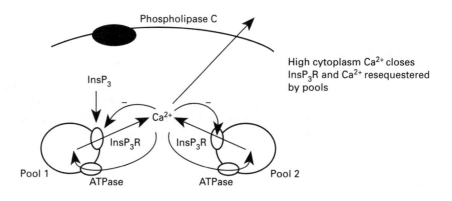

High cytoplasm Ca^{2+} closes $InsP_3R$ and Ca^{2+} resequestered by pools

posed that it has a second-messenger role in many cells. The effect of cADPR on RyRs and therefore Ca^{2+} release can be inhibited by the presence of either ruthenium red or 8-amino-cADPR. Levels of cADPR might well be modulated by cGMP in some instances, suggesting a control mechanism that might be physiologically important. cGMP is the product of guanylyl cyclase, which itself might be under the control of the gas NO. NO might be produced by the same cell or produced by a neighbouring cell. The production and roles of NO are discussed further in Chapter 8.

Nicotinate Adenine Dinucleotide Phosphate

Recently, nicotinate adenine dinucleotide phosphate ($NAADP^+$), a deaminated derivative of nicotinamide adenine dinucleotide phosphate ($NADP^+$), has been shown to have the capacity to release Ca^{2+} from intracellular stores. This activity is independent of the release triggered by $InsP_3$ and cADPR. The release triggered by $NAADP^+$ is not inhibited by heparin, which interferes with $InsP_3$ responses, and is not affected by the presence of ruthenium red, used to block the cADPR response. Therefore such results suggest a separate pathway that a cell could use for signalling the release of Ca^{2+} ions. The response is selectively inhibited by the presence of thionicotinamide-$NADP^+$, which will no doubt be of great use in the elucidation of the exact role of $NAADP^+$ in Ca^{2+} signalling in many cells.

Apoptosis

One of the puzzles of modern biochemistry is the observation that some cells seem to undergo a process of programmed death, a process called apoptosis. Prolonged rises in Ca^{2+} concentrations in thymocytes cause apoptosis and this suggests that the intracellular Ca^{2+} concentration is vital in this phenomenon. It was suggested that a Ca^{2+}/Mg^{2+}-dependent endonuclease causes double-stranded cleavage of DNA, leading to cell death. More recent research has suggested a role for calmodulin-dependent enzymes as well as Ca^{2+}-buffering proteins. Calmodulin may be having some of its apoptotic effects through activation of the cAMP pathways. One buffering protein singled out as important is a vitamin D-dependent Ca^{2+}-binding protein called calbindin D_{28K}. Overexpression of this protein in lymphocytes seems to protect against apoptosis induced by several factors, including Ca^{2+} ionophores and glucocorticoids.

Sustained rises in cytoplasmic Ca^{2+} ion concentrations may lead to stimulation of the activities of endonucleases, proteases, kinases, phosphatases and phospholipases, all of which may lead to a rapid loss of cell function and integrity. Such mechanisms, amongst others, have been suggested to play a major role in this phenomenon.

Fluorescence Detection and Confocal Microscopy

Methods of study of Ca^{2+} control have often relied on the presence of reporter molecules. One such probe is the photoprotein aequorin. This protein, which emits light on the binding of Ca^{2+}, is purified from a luminous jellyfish (*Aequorea forskalea*) and used to sense Ca^{2+} concentrations in the nanomolar to micromolar range.

The study of the intracellular concentrations of Ca^{2+} and its role in signalling in the cell has recently been greatly enhanced by the use of fluorescent probes. These chemicals all possess the ability to bind to free Ca^{2+} ions and, on binding, their fluorescent spectral characteristics change, which can then be readily detected by several techniques. Such probes include Fura-2 and Fluo-3. The absorbance change may be simply detected on a gross scale by the use of a fluorimeter, which monitors the change in emitted light if the sample is excited at a preset wavelength of light, fixed or scanning; however, this technique is limited in that a population of cells is required to obtain a detectable signal. A fluorescence microscope overcomes this problem, as it can be focused on single cells; interestingly, such studies often show that two cells from the same tissue treated with the same stimulant may not necessarily respond in the same way. An even more powerful technique is the use of a confocal scanning laser microscope. Here, the light source is from a laser as opposed to a normal white light source, and the laser is driven to scan in the X–Y plane. The laser is focused on to the sample, enabling a single image to be obtained from a very small area, often a single cell; more excitingly, the fluorescence emission is also focused back to the detector, usually a photomultiplier tube, which sends a signal to a computer that creates an image on a VDU screen. The inherently grey image can then be enhanced by the use of pseudo-colour imaging, enabling the user to obtain an easy-to-see scale of fluorescence in the field of view. The power of this type of system comes from the fact that the laser not only scans in the horizontal plane but also is extremely finely focused in the vertical plane. Therefore, single cells, or indeed whole tissues, can be optically sectioned, allowing a three-dimensional image to be created. These relatively new techniques have led to great advances in the study of the role of intracellular Ca^{2+}. The concentration of Ca^{2+} ions inside organelles can now be visualised, giving an insight into the changes in these concentrations in, for example, the nucleus. New roles for Ca^{2+} signalling, such as in the control of gene expression, will undoubtedly come to light as the use of such equipment expands.

Summary

Ca^{2+} ions are unusual signalling molecules in that they are neither created nor destroyed; the signal comes from a sudden change in their distribution. Cells usually expend large amounts of energy actively removing Ca^{2+} ions, using pumps in their plasma membrane. These maintain a gradient, with the intracellular Ca^{2+} concentration at less than 10^{-7} mol/l, and the extracellular Ca^{2+} concentration in the region of 2×10^{-3} mol/l. Pumps in the ER, sarcoplasmic reticulum and mitochondria are also used to sequester Ca^{2+} into intracellular stores. A signal arises when the cytoplasmic Ca^{2+} concentration is allowed to suddenly rise, often by the release of Ca^{2+} from intracellular stores, which is commonly mediated by the presence of $InsP_3$ derived from the breakdown of lipids on the plasma membrane, an event which itself is tightly controlled (Figure 7.11).

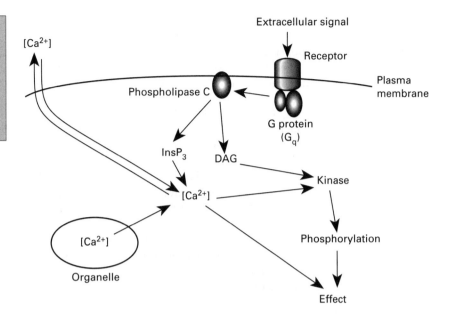

Figure 7.11
Simplified scheme showing the control and role of Ca^{2+} ion concentrations within the cell. DAG, diacylglycerol; $InsP_3$, inositol trisphosphate.

As well as the actual presence of Ca^{2+} ions, the signal commonly relies on the presence of Ca^{2+}-binding proteins. These quite commonly contain polypeptide domains known as EF hands. The most ubiquitous is calmodulin. This protein has two Ca^{2+}-binding domains, each capable of binding two Ca^{2+} ions, connected by a flexible α-helix.

Ca^{2+} is known to control many cell functions, including regulation of PKC, Ca^{2+}/calmodulin-dependent protein kinase, cell cycles, metabolic enzymes and even other signalling routes, such as nitric oxide synthase.

Recent research, particularly with the use of fluorescent probes and confocal microscopy, has revealed that the concentration of Ca^{2+} ions in the cytoplasm is not uniform in nature. Ca^{2+} ions are seen to have localised concentrations, to move in waves and to undergo rapid oscillations in the measured concentration. Such phenomena are a challenge to the full understanding of how Ca^{2+} ions coordinate the control of cellular functions, especially in events such as programmed cell death or apoptosis.

Further Reading

Berridge, M. J. (1993) Inositol trisphosphate and calcium signalling. *Nature*, **361**, 315–25.

Chao, C. P., Laulederkind, S. J. F. and Ballou, L. R. (1994) Sphingosine-mediated phosphatidylinositol metabolism and calcium mobilisation. *Journal of Biological Chemistry*, **269**, 5849–56.

Clapham, D. E. (1995) Calcium signaling. *Cell*, **80**, 259–68.

Clapham, D. E. (1995) Replenishing stores. *Nature*, **375**, 634–5.

Clapham, D. E. and Sneyd, J. (1995) Intracellular calcium waves. *Advances in Second Messenger and Phosphoprotein Research*, **30** 1–24.

Cooper, D. M. F., Mons, N. and Karpen, J. W. (1995) Adenylyl cyclases and the interaction between calcium and cAMP signalling. *Nature*, **374**, 421–4.

Kretsinger, R. H. (1980) Structure amd evolution of calcium modulated proteins. *CRC Critical Reviews in Biochemistry*, **8**, 119–74.

Lorca, T., Cruzalegui, F. H., Fesquet, D., *et al.* (1993) Calmodulin-dependent protein kinase II mediates inactivation of MPF and CSF upon fertilisation of *Xenopus* eggs. *Nature*, **366**, 270–3.

Mattie, M., Brooker, G. and Spiegel, S. (1994) Sphingosine 1-phosphate, a putative second messenger, mobilizes calcium from internal stores via an inositol trisphosphate-independent pathway., *Journal of Biological Chemistry*, **269**, 3181–8.

Means, A. R., (1994) Calcium, calmodulin and cell-cycle regulation. *FEBS Letters*, **347**, 1–4.

Nayal, M. and Di Cera, E. (1994) Predicting Ca^{2+} binding sites in proteins. *Proceedings of the National Academy of Sciences USA*, **91**, 817–21

Negulescu, P. A., Shatri, N. and Cahalan, M. D. (1994) Intracellular calcium dependence on gene expression in single T-lymphocytes. *Proceedings of the National Academy of Sciences USA*, **91**, 2873–7.

Petersen, O. H., Petersen, C. H. and Kasai, H. (1994) Calcium and hormone action. *Annual Review of Physiology*, **56**, 297–319.

Tsien, R.W. and Tsien, R.Y. (1991) Calcium channels, stores and oscillations. *Annual Review of Cell Biology*, **6**, 715–60.

Calmodulin

Babu, Y., Sack, J. S., Greenhough, T. J., Bugg, C. E., Means, A. R. and Cook, W. J. (1985) Three-dimensional structure of calmodulin. *Nature*, **315**, 37–40.

Barbato, G., Ikura, M., Kay, L. E., Pastor, R. W. and Bax, A. (1992) Backbone dynamics of calmodulin studies by N-15 relaxation using inverse detected two-dimensional NMR spectroscopy: the central helix is flexible. *Biochemistry*, **31**, 5269–78.

Billingsley, M. L., Polli, J. W., Pennypacker, K. R. and Kincaid, R. L. (1990) Identification of calmodulin-binding proteins. *Methods in Enzymology*, **184**, 451–67.

Finn, B. E. and Forsen, S. (1995) The evolving model of calmodulin structure, function and activation. *Structure*, **3**, 7–11.

Head, J. F., (1992) A better grip on calmodulin. *Current Biology*, **2**, 609–11.

James, P., Vorherr, T. and Carafoli, E. (1995) Calmodulin-binding domains: just two faced or multi-faceted. *Trends in Biochemical Sciences*, **20**, 38–42.

O'Neil, K. T. and DeGrado, W .F. (1990) How calmodulin binds its targets: sequence independent recognition of amphipathic α-helices. *Trends in Biochemical Sciences*, **15**, 59–64.

Quadroni, M., James, P. and Carafoli, E. (1994) Isolation of phosphorylated calmodulin from rat liver and identification of the *in vivo* phosphorylation sites. *Journal of Biological Chemistry*, **269**, 16116–22.

Takuwa, N., Zhou, W. and Takuwa, Y. (1995) Calcium, calmodulin and cell-cycle progression. *Cellular Signalling*, **7**, 93–104.

The Plasma Membrane and its Role

Pederson, P. L. and Carafoli, E. (1987) Ion motive ATPases. 1: Ubiquity, properties, and significance to cell function. *Trends in Biochemical Sciences*, **12**, 146–50.

Phillipp, S., Cavalié, M. F., Wissenbach, U., *et al.* (1996) A mammalian capacitative calcium entry channel homologous to *Drosophila* TRP and TRPL. *EMBO Journal*, **15**, 6166–71.

Intracellular Stores

Finch, E. A. and Goldin, S. M. (1994) Calcium and inositol 1,4,5-trisphosphate-induced Ca^{2+} release. *Science*, **265**, 813–15.

MacLennan D. H., Toyofuku, T. and Lytton, J. (1992) Structure–function relationships in sarcoplasmic or endoplasmic reticulum type Ca^{2+} pumps. *Annals of the New York Academy of Sciences*, **671**, 1–10.

Meissner, G. (1994) Ryanodine receptor/Ca^{2+} release channels and their regulation by endogenous effectors. *Annual Review of Physiology*, **56**, 485–508.

Mignery, G. A. (1990) The ligand binding site and transduction mechanism in the inositol-1,4,5-trisphosphate receptor. *EMBO Journal*, **9**, 3893–8.

Pozzan, T., Rizzulo, R., Volpe, P. and Meldolesi, J. (1994) Molecular and cellular physiology of intracellular calcium stores. *Physiological Reviews*, **74**, 595–636.

Stehno-Bittel, L., Lückhoff, A. and Clapham D. E. (1995) Calcium release from the nucleus by $InsP_3$ receptor channels. *Neuron*, **14**, 163–7.

Toyoshima, C., Sassabe, H. and Stokes, D. L. (1993) Three-dimensional cryo-electron microscopy of the calcium ion pump in the sarcoplasmic reticulum membrane. *Nature*, **362**, 469–71.

Gradients, Waves and Oscillations

Allbritton, N. L. and Meyer, T. (1993) Localised calcium spikes and propagating calcium waves. *Cell Calcium*, **14**, 691–7.

Meyer, T. and Stryer, L. (1991) Calcium spiking. *Annual Review of Biophysics and Biophysical Chemistry*, **20**, 153–74.

Rapp, P. E. and Berridge, M. J. (1981) The control of transepithelial potential oscillations in the salivary gland of *Calliphora erythrocephala*. *Journal of Experimental Biology*, **93**, 119–32.

Tsunoda, Y. (1991) Oscillatory Ca^{2+} signalling and its cellular function. *New Biologist*, **3**, 3–17.

Sphingosine-1-Phosphate.

Chao, C. P., Laulederkind, S. J. F. and Ballou, L. R. (1994) Sphingosine mediated phosphatidylinositol metabolism and calcium mobilization. *Journal of Biological Chemistry*, **269**, 5849–56.

Ghosh, T.K., Bian, J. and Gill, D.L. (1994) Sphingosine 1-phosphate generated in the endoplasmic reticulum membrane activates release of stored calcium. *Journal of Biological Chemistry*, **269**, 22628–35.

165

Mattie, M., Brooker, G. and Spiegel, S. (1994) Sphingosine 1-phosphate, a putative second messenger, mobilizes calcium from internal stores via an inositol trisphosphate independent pathway. *Journal of Biological Chemistry*, **269**, 3181–8.

Yule, D. I., Wu, D., Essington, T. E., Shayman, J. A. and Williams, J. A. (1993) Sphingosine metabolism induces calcium oscillations in rat pancreatic acinar cells. *Journal of Biological Chemistry*, **268**, 12353–8.

Cylic ADP Ribose

Berridge, M. J. (1993) A tale of two messengers. *Nature*, **365**, 388–9.

Gallione, A. (1993) Cyclic ADP-Ribose: a new way to control calcium. *Science*, **259**, 325–6.

Mészáros, L. G., Bak, J. and Chu, A. (1993) Cyclic ADP-ribose as an endogenous regulator of the non-skeletal type ryanodine receptor Ca^{2+} channel. *Nature*, **364**, 76–9.

Thorn, P., Gerasimenko, O. and Petersen, O.H. (1994) Cyclic ADP-ribose regulation of ryanodine receptors involved in agonist-evoked cytosolic Ca^{2+} oscillations in pancreatic acinar cells. *EMBO Journal*, **13**, 2038–43.

Nicotinate Adenine Dinucleotide Phosphate

Aarhus, R., Dickey, D.M., Graff, R.M., Gee, K.R., Walseth, T.F. and Lee, H.C. (1996) Activation and inactivation of Ca^{2+} release by $NAADP^+$. *Journal of Biological Chemistry*, **271**, 8513–16.

Chini, E.N., Beers, K.W. and Dousa, T.P. (1995) Nicotinate adenine-dinucleotide phosphate (NAADP) triggers a specific calcium release system in sea urchin eggs. *Journal of Biological Chemistry*, **270**, 3216-23.

Apoptosis

Harman, A. W. and Maxwell, M. J. (1995) An evaluation of the role of calcium in cell injury. *Annual Review of Pharmacology and Toxicology*, **35**, 129–44.

Lee, S., Christakos, S. and Small, M.B. (1993) Apoptosis and signal transduction: clues to a molecular mechanism. *Current Opinion in Cell Biology*, **5**, 286–91.

Nicotera, P., Zhivotovsky, B. and Orrenius, S. (1994) Nuclear calcium transport and the role of calcium in apoptosis. *Cell Calcium*, **16**, 279–88.

Confocal Microscopy

Rizzuto, R., Brini, M., Murgia, M. and Pozzan, T. (1993) Microdomains with high Ca^{2+} close to $InsP_3$-sensitive channels that are sensed by neighbouring mitochondria. *Science*, **262**, 744–7.

Williams, D. A. (1993) Mechanisms of calcium release and propagation in cardiac cells: do studies with confocal microscopy add to our understanding? *Cell Calcium*, **14**, 724–35.

8

Nitric Oxide, Hydrogen Peroxide and Carbon Monoxide

Topics

- Nitric Oxide
- Superoxide and Hydrogen Peroxide
- Carbon Monoxide

Introduction

In recent years, particularly since the realisation that endothelium-derived relaxing factor (EDRF) is in fact NO, it has become apparent that a new and, at first glance, rather surprising group of molecules are involved in cell signalling. These molecules include NO, hydrogen peroxide (H_2O_2) and carbon monoxide (CO). On the face of it, these compounds do not appear to have the right properties to be good signalling molecules and, moreover, they have the potential to be detrimental to the cells in which they are formed and to the cells around them. However, research papers continue to appear regularly in the literature citing ever-increasing experimental evidence for the roles of these molecules in the control of cellular functions, including apparent direct effects on gene transcription. NO, for example, regulates guanylyl cyclase activity, controls neurotransmitter release, acts as a neurotransmitter, and has bactericidal and tumoricidal activities.

H_2O_2, NO and CO are all characterised by being small inorganic molecules. Generally they are extremely reactive and have known activity towards biological materials, often causing alteration and loss of function. Molecules such as superoxide and H_2O_2 have well-established classical functions in the body, i.e. the destruction of invading organisms as a major part of host defence. In fact, this is not unique to the animal kingdom, with more and more evidence now showing a similar role for superoxide and H_2O_2 in plants, possibly leading to local cell death and areas of necrosis. However, despite this, these molecules appear to be produced by many non-phagocytic cells of animals, as well as in plants, where their role may be to control cellular events including the rate of cell proliferation.

Nitric Oxide

The small gaseous molecule NO has been recently discovered to be a significant signalling molecule. It is not restricted to one particular tissue but has functions in various and diverse locations. Originally described as a factor which caused the relaxation of cells (EDRF), in 1987 Moncada and colleagues realised that this activity was mediated by NO. Furthermore, it was found that some treatments for angina, for example nitroglycerine, are mediated through NO. Nitroglycerine or other organic nitrates are converted to NO, which causes relaxation of the blood vessels, hence increasing the heart's blood supply. Macrophages appear to use NO in part of their host-defence mechanism, while some functions of brain and other nervous tissue are inhibited by compounds that reduce the production of NO.

NO is a free radical, commonly written as NO$^{\bullet}$, i.e. it contains an unpaired electron, denoted by the superscript dot in the chemical notation. This leads to its increased reactivity because this is an unfavoured electronic state and one from which a molecule is keen to escape. NO is produced by the enzyme nitric oxide synthase (NOS). The effects of NO are usually local, owing to its short half-life, in the order of only 5–10 s. However it can readily diffuse across membranes, which means that its signalling action is not restricted to the cell of origin and neighbouring cells are also affected. The signal is usually turned off by the deactivation of NO, which is converted to nitrates and nitrites by oxygen and water.

NO is formed by the oxidation of L-arginine (Figure 8.1). The guanidine group

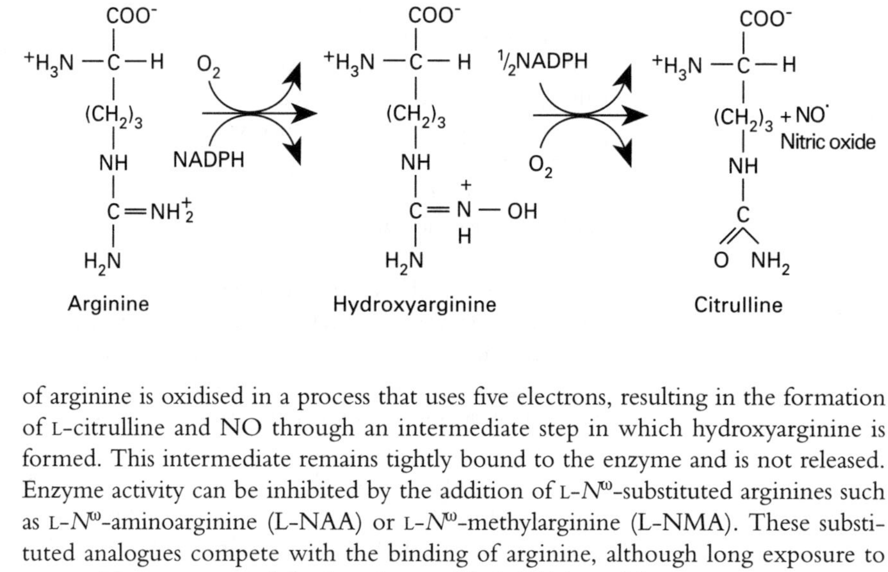

Figure 8.1
Production of nitric oxide from arginine catalysed by nitric oxide synthase.

of arginine is oxidised in a process that uses five electrons, resulting in the formation of L-citrulline and NO through an intermediate step in which hydroxyarginine is formed. This intermediate remains tightly bound to the enzyme and is not released. Enzyme activity can be inhibited by the addition of L-N^{ω}-substituted arginines such as L-N^{ω}-aminoarginine (L-NAA) or L-N^{ω}-methylarginine (L-NMA). These substituted analogues compete with the binding of arginine, although long exposure to some of these compounds leads to irreversible inhibition of the enzyme.

The enzyme responsible for the production of NO, NOS, is not a single enzyme but rather a family of isoforms. Using the observation that calmodulin was required for activation of the enzyme, Bredt and Snyder first purified the enzyme from brain tissue in 1990. This was denoted bNOS, a protein of approximately 160 kDa. This led to the purification of isoforms from macrophages (macNOS) and endothelial cells (eNOS; approximately 133 kDa). Subsequent Southern blot analysis has suggested that mammalian genomes contain three genes encoding for NOS isoforms.

Two of these are expressed constitutively, an endothelial-type enzyme (eNOS) and a neuronal type (nNOS), the latter having an expression pattern that includes non-neuronal tissues. The third gene encodes an inducible form (iNOS), which includes the enzyme seen in macrophages. Furthermore, some forms of the enzyme are seen as dimers *in vivo* and it has been suggested that some forms may exist in an equilibrium between monomeric and dimeric states.

Cloning of the NOS isoenzymes has revealed that they all share close homology to the enzyme cytochrome P450 reductase (Figure 8.2). In this enzyme, an NADPH

Figure 8.2
Domain structure of isoforms of nitric oxide synthase (NOS) with the area of homology to cytochrome P450 reductase highlighted.

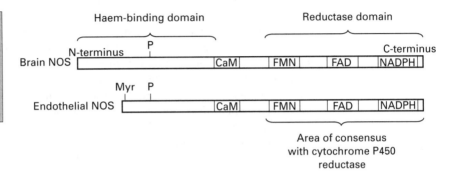

binding site has been identified, as well as areas for flavin adenine dinucleotide (FAD) and flavin mononucleotide (FMN) binding. Comparison of the two enzyme sequences reveals consensus sequences for the binding of these cofactors. Purification of NOS has shown that each monomer has one FAD and one FMN although, like other enzymes, FAD slowly dissociates and has to be exogenously added to obtain full activity. The similarity to the cytochrome P450 system continues in that NOS also contains a haem prosthetic group, i.e. iron protoporphyrin IX, as in guanylyl cyclase (see Chapter 5). Like many haem-containing enzymes NOS reacts with and is inhibited by CO, giving a characteristic CO-binding absorbance spectrum which suggests that the haem is attached via a cysteine residue. Such binding of CO is characteristic of enzymes that have the capacity to bind to oxygen, as seen with cytochrome oxidase and haemoglobin. It appears that the first step of the catalysis is the binding of the arginine to the haem group, with subsequent oxidation reactions.

Between the FMN and FAD binding regions and the haem-binding region is a region used for calmodulin binding. At the N-terminal end of this particular part of the polypeptide is a trypsin-sensitive region. Digestion with trypsin yields two domains, one from the N-terminal end of NOS that binds to arginine and contains haem, and a second containing the NADPH- and flavin-binding regions. It has been suggested therefore that the synthase is a bi-domain enzyme, with one domain containing the oxygenase activity and the other domain containing the reductase activity, where the two domains may even be able to function independently of each other. However, the exact stoichiometry of the electron transfer has yet to be fully determined, with electron leakage possible to molecular oxygen that would yield the superoxide ion, also an unstable free radical (see below).

Purification of the brain enzyme showed that tetrahydrobiopterin is tightly bound to NOS and it was thought that it must function in the catalysis as it could take part in an electron-transfer role. However, this is probably not the case and it is likely that tetrahydrobiopterin acts only in a stabilising role.

Not only is the binding of calmodulin to the enzyme crucial to its original purification but it has become apparent that Ca^{2+} is important in the regulation of many isoforms of NOS. The concentrations of Ca^{2+} involved are in the region to be expected for a calmodulin-activated system, with an EC_{50} of $2-4 \times 10^{-7}$M. Neither Ca^{2+} nor calmodulin appear to affect arginine binding but calmodulin binding seems to regulate the electron-transfer activities of the enzyme.

As mentioned above, whereas some NOSs are produced by cells constitutively other forms are inducible, including NOS of macrophages, designated mNOS or iNOS. This is a synthase of approximately 130 kDa and exists as a dimer of two identical subunits, i.e. a homodimer. It is interesting to note that these isoforms contain calmodulin-binding sites but are in fact unaffected by calmodulin antagonists or Ca^{2+}; indeed calmodulin is tightly bound to these enzymes even in the absence of Ca^{2+}. Production of new NOS protein molecules of these isoforms is stimulated by the presence of, amongst many others, IFN-γ, IL-1 and lipopolysaccharide (LPS). Cloning of the transcription start site of the NOS gene has shown that two distinct regions contain LPS- and IFN-responsive elements. The LPS region lies about 50–200 base pairs upstream of the transcription start site while the region responsive to IFN-γ is about 900–1000 base pairs upstream from the start site. Other response elements have also been reported. Furthermore, cells other than macrophages have also been shown to contain the inducible form of the enzyme.

Consensus sequences for phosphorylation by cAPK have been found in the NOS of brain and endothelial cells, although the macrophage form of the enzyme seems to lack them. cAPK, PKC, cGPK and Ca^{2+}/calmodulin-dependent protein kinase can all phosphorylate the neuronal form of the enzyme resulting in a decrease in NOS activity.

It is not only the enzymatic activity that is regulated by phosphorylation but also the subcellular distribution of the enzyme in endothelial cells. The NOS of these cells is primarily located in the plasma membrane. On addition of bradykinin, the enzyme is phosphorylated and is translocated to the soluble fraction of the cells, albeit in an inactive state. This mechanism would ensure that the enzyme is only active when it is located where NO is released from the cell, i.e. when it is associated with the plasma membrane. Enzyme location may also be influenced by other factors. For example, the eNOS enzyme is myristoylated at the N-terminal end; if this site is removed by site-directed mutagenesis the enzyme alters its location from the membrane to the cytosol. There are also reports that the NOS polypeptide may be palmitoylated.

But what does NO actually do? It is certain that NO acts on the enzyme guanylyl cyclase, the enzyme responsible for the production of cGMP, itself a second messenger (see Chapter 5). The activation of guanylyl cyclase is caused by the binding of NO to the haem group of the enzyme. In some brain tissues it appears that NO is responsible for the regulation of the production of cGMP through activation of guanylyl cyclase; cGMP itself might be responsible for the regulation of serine/threonine kinases and may regulate the activity of some cAMP phosphodi-esterases, with resultant modulation of any cAMP response (see Chapter 5). In vascular tissues the activation of guanylyl cyclase by NO causes muscle relaxation, probably by the activation of a cGPK acting on myosin light chains.

NO is also able to bind iron not associated with haem groups in some enzymes, for example, enzymes containing iron–sulphur complexes such as NADH-ubiquinone oxidoreductase, otherwise known as complex I of the mitochondrial electron transport chain. Its interaction with such non-haem iron has been implicated in regulation of the translation of some proteins and even in the alteration of rates of DNA synthesis. Furthermore, the binding of NO to ferritin causes the liberation of free iron, which in the presence of oxygen free radicals leads to lipid peroxidation of membranes. This would be a very detrimental activity in the cell.

In neuronal tissues, NO is involved in the regulation of neurotransmitter release. Interestingly, both the differentiation and regeneration of neurones also might be affected by NO and it has been reported that brain NOS is transiently expressed after neuronal injury. NO also acts as a specific neurotransmitter. Examples of this can be seen in the relaxation of the smooth muscles involved in peristalsis as well as in the control of penile erections. However, if the NO concentration rises above the normally very low levels, neurotoxicity can result. NO reacts with another free radical, superoxide, itself a possible product from NOS. This results in the production of a very reactive species, peroxynitrite, which potentially causes cellular damage.

Besides its role as a cell-signalling molecule, like superoxide and its by-products NO also has tumoricidal and bactericidal activities and it has been reported that viral replication is also inhibited by NO.

Superoxide and Hydrogen Peroxide

The early discoveries in 1933 by Baldridge and Gerard that oxygen consumption increased in stimulated phagocytic cells has led to great interest in the molecular events which take place on the activation of these cells. Sbarra and Karnovsky in 1959 overturned the earlier view that the oxygen was used for increased respiration, and by 1961 Iyer and his group noted that oxygen consumption was accompanied by an increase in the hexose monophosphate shunt leading to the production of NADPH. It is now known that oxygen is in fact used in the production of superoxide ions, by the direct enzymatic reduction of the oxygen molecule by a complex situated in the plasma membrane of phagocytic cells. This enzyme complex is known as the NADPH oxidase and has been well characterised in neutrophils. The importance of this biological activity is clearly demonstrated by the disease chronic granulomatous disease (CGD). This is a genetic disease that manifests itself as a defect of the NADPH oxidase complex with the complete abolition of any superoxide production. People with this disorder suffer from recurrent bacterial and fungal infections and until relatively recently died at an early age.

The primary product from the NADPH oxidase enzyme is the superoxide ion, where the electrons are supplied by intracellular NADPH:

$$2O_2 + NADPH \rightarrow 2O_2^- + NADP^+ + H^+$$
$$\text{superoxide ion}$$

The extra electron supplied to molecular oxygen is in an unpaired state and hence the ion is classified as a free radical. This electronic state is unstable and the new ion is consequently very reactive. It readily undergoes dismutation with the formation of H_2O_2, not itself a free radical but often grouped with the oxygen free radicals:

$$2O_2^- + 2H^+ \rightarrow H_2O_2 + O_2$$
$$\text{hydrogen peroxide}$$

Although this reaction occurs spontaneously, especially at low pH, it is also catalysed by an enzyme called superoxide dismutase. This enzyme has two main forms: a copper/zinc containing form that resides in the cytosol of cells and a manganese-containing form located in mitochondria.

Both superoxide ions and H_2O_2 are reactive towards biological materials, although it is probably the results of a further cascade of reactions that causes the most damage. Superoxide causes biological oxidation especially in hydrophobic environments, while H_2O_2, a relatively weak oxidising agent, inactivates some enzymes, usually by the oxidation of thiol groups. Like the removal of superoxide by superoxide dismutase, H_2O_2 is catalytically destroyed by either the enzyme catalase or the glutathione cycle. The importance of the abolition of superoxide is demonstrated by the fact that all aerobic organisms appear to contain at least one isoform of superoxide dismutase, life in the presence of oxygen being dependent on the destruction of harmful oxygen free radicals. However, the real danger comes when superoxide and H_2O_2 react together with formation of hydroxyl radicals:

$$O_2^- + H_2O_2 \rightarrow OH^\bullet + OH^- + O_2$$
$$\text{hydroxyl radical}$$

This reaction is catalysed by the presence of iron ions, Fe^{2+}, or copper ions, Cu^{2+}, proceeding via what is known as the Haber–Weiss reaction. Hydroxyl radicals are extremely reactive, the free unpaired electron making it extremely unstable. Cell damage is caused by oxidation of proteins, oxidation of bases, which can lead to DNA strand breakage, direct attack on deoxyribose moieties and lipid peroxidation, as mentioned above. Usually such reactions proceed via removal of hydrogen atoms from organic molecules, often leading to the formation of new radicals and possibly a cascade of further free-radical reactions. In the case of lipid peroxidation this can lead to membrane dysfunction.

Other oxygen free radicals are also formed as a consequence of the production of superoxide. These include singlet oxygen, which can lead to the destruction of carotenes, haem proteins and membrane lipids and, in phagocytic cells where the haem-containing enzyme myeloperoxidase is present, the production of hypochlorite.

Interestingly, superoxide can also react with NO to produce a very reactive compound, peroxynitrite.

$$NO^{\bullet} + O_2^- \rightarrow OONO^-$$
$$\text{peroxynitrite}$$

Therefore, it is possible that the production of superoxide ions modulates the role of NO, albeit with the production of a very reactive compound.

Despite this apparent cascade of dangerous chemicals produced as a consequence of the release of superoxide, both superoxide ions and H_2O_2 appear to be released in non-phagocytic cells. Here the function is almost certainly not in host defence. The levels of superoxide are only about 1% of those in neutrophils, while production is sustained for very long periods of time, possibly constitutively by some cells. Different cell types might use these free radicals in different ways but several lines of evidence point to a cell-signalling role in many instances.

NADPH Oxidase

One of the most well-characterised sources of superoxide, and hence other oxygen free radicals, is from the enzyme NADPH oxidase. It is in fact an enzyme complex, catalysing the one-electron reduction of molecular oxygen to superoxide. The preferred electron donor is NADPH (K_m approximately 50 mM), and not NADH (K_m approximately 500 mM) and hence the complex gets its name.

NADPH oxidase is a membrane-bound complex consisting of a short electron transport chain, containing only two redox active groups, an FAD moiety and a cytochrome (Figure 8.3). The cytochrome involved is a b-type cytochrome, located in the plasma membrane and specific granules of neutrophils. However, more recently it has been found in the plasma membrane of several other cell types including other phagocytes such as macrophages and also non-phagocytic cells such as fibroblasts and mesangial cells from the kidney. The cytochrome is unique in that it has a very low mid-point potential, measured as $E_{m7.0} = -245$ mV, considerably lower than that of other eukaryotic cytochromes. However, this extremely low mid-point potential is sufficiently low to facilitate the reduction of molecular oxygen to superoxide and it has also been shown that the cytochrome is kinetically competent to act as the direct electron donor to oxygen. Because of this unique characteristic, the cytochrome is referred to as cytochrome b_{-245} but commonly is known by the wavelength of its α-band absorbance in the visible spectrum, i.e. cytochrome b_{558}.

The cytochrome is also unusual in being a heterodimer, containing a small α-subunit of approximate molecular mass of 22 kDa. The protein has been named p22-*phox*, the *phox* referring to the fact that is was characterised from a phagocytic cell oxidase. The larger ß-subunit of the cytochrome has an approximate molecular mass of 76–92 kDa, and is highly glycosylated. It is referred to as gp91-*phox*. Radiation inactivation experiments have suggested that the haem prosthetic group may be attached to the smaller of the two subunits, the p22-*phox* subunits, while other researchers have claimed to have purified the small subunit with the haem attached. However electron paramagnetic resonance and Raman spectroscopy indicates that the haem is low spin, six coordinate with imidazole-like ligands, while sequencing

Figure 8.3
Schematic representation of the NADPH oxidase complex and its activation by translocation of several cytosolic components to the plasma membrane.

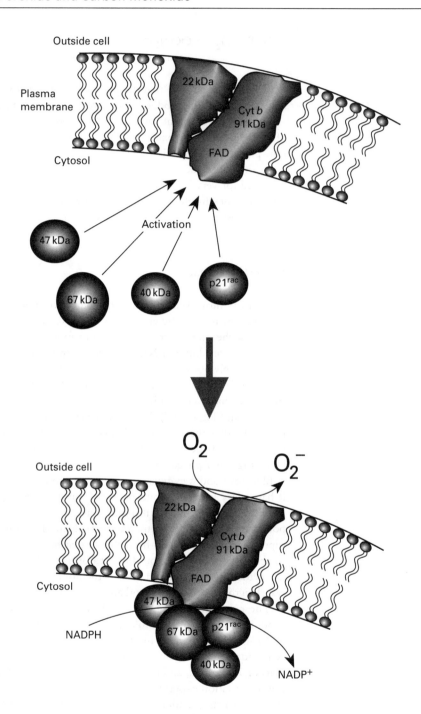

data shows that the small subunit contains only one invariant histidine residue. The other histidine of the protein is polymorphic and therefore probably not involved in haem binding. Consequently, the exact nature of haem binding remains unknown, and it has been suggested that the cytochrome has two haem groups of slightly different redox potentials.

The large subunit of the cytochrome b, gp91-*phox*, is also believed to contain several membrane-spanning domains. Although, like haem binding, the exact site of FAD binding has been contested, with several putative polypeptides being suggested, it has recently been shown that the large cytochrome b subunit, gp91-*phox*, contains a site of FAD attachment and probably acts as the flavoprotein dehydrogenase. Both cytochrome subunits have been cloned and sequenced and strong homology exists between the amino acid sequence of the large subunit and that of several ferredoxin-$NADP^+$ reductases, showing conservation in the nucleotide-binding regions. Also, activity has been reconstituted using purified proteins, where the cytochrome b is the only one to be assigned an electron-transfer activity. Using electron spin resonance, the mid-point potential of the FAD has been found to be $-280\,mV$, low enough to be thermodynamically capable of participating in the reactions.

Besides the membrane-bound cytochrome b_{-245}, the oxidase also requires the presence of several cytosolic proteins (see Figure 8.3). By studying patients with CGD, two polypeptides have been identified as being integral with NADPH oxidase activity. These have molecular masses of 47 kDa (p47-*phox*) and 67 kDa (p67-*phox*). Both these proteins are primarily found in the cytosol of resting neutrophils but on activation are translocated to the plasma membrane. Both have been cloned and sequenced and both contain SH3 domains, which are probably used in their interaction with the other polypeptides in the complex. Although p47-*phox* is phosphorylated on activation, this does not appear to be a prerequisite for full activity and as yet neither of these cytosolic proteins have been assigned specific functions. Interestingly, these polypeptides are also expressed in non-myeloid cells, as determined by the use of reverse-transcription polymerase chain reaction (PCR) and antibody-binding studies.

Recently, another cytosolic NADPH oxidase component containing an SH3 domain has been recognised. This is a 40 kDa polypeptide, p40-*phox*, that appears to form an activation complex with p47-*phox* and p67-*phox*, which as a unit translocates to the plasma membrane to associate with the flavocytochrome b. Sequence data reveals that this polypeptide shares a large region of homology with the N-terminus of p47-*phox*.

Other cytosolic proteins required for full activity of the NADPH oxidase were found by the reconstitution of activity from mixtures of cytosol and membranes. Besides p47-*phox* and p67-*phox* it was found that a heterodimeric complex is required. This complex contained a G protein related to Ras, $p21^{rac}$, and a GDP-dissociation inhibition factor, Rho-GDI. These two proteins undergo a cycle of dissociation and association with the subsequent turnover of GTP (discussed more fully in Chapter 5).

Through the action of the NADPH oxidase, accepting an electron on the inside of the membrane and giving it up on the outside to oxygen, the enzyme system probably works in an electrogenic manner and a compensatory movement of H^+ through an appropriate channel has been suggested. However, the exact nature of this channel has yet to be identified, although it has been suggested that the movement of H^+ is vital to the killing process carried out by phagocytic cells.

175

Chronic Granulomatous Disease

To fully understand the role of oxygen free radicals in the body it is necessary to have a brief discussion of CGD. This is a rare inherited disorder characterised by a lack of production of oxygen free radicals by neutrophils; the molecular defects have been defined as a lack of NADPH oxidase activity of these cells. This leads to a reduction in host defence with the patient suffering recurrent bacterial and fungal infections, leading to death if untreated.

Early studies on the molecular defects occurring in CGD showed that a low-potential cytochrome b was missing from the neutrophils of four patients, although subsequent studies in Denmark showed that some patients with the disease had normal amounts of cytochrome b_{-245}. A multicentre European evaluation of the incidence of CGD was mounted where the cytochrome was measured in 27 patients with CGD and 64 members of their families. The cytochrome b was undetectable in all 19 men who carried the defect on their X chromosome. Heterozygous carriers appeared to have reduced concentrations of the cytochrome and, interestingly, variable proportions of their neutrophils were unable to release superoxide. In all eight patients where inheritance was probably autosomal the cytochrome was present but not functional.

It is now apparent that inheritance of CGD can be both X-linked and autosomal, with defects assigned to one of the four main components of the NADPH oxidase: gp91-*phox*, p22-*phox*, p47-*phox*, p67-*phox*.

The most common form of the disease is X-linked, accounting for approximately 50% of patients. This is usually seen as a complete lack of the activity of the gp91-*phox* cytochrome subunit. However, it has been reported that both subunits of the cytochrome are missing in this form of the disease as both subunits are needed for stable incorporation of the cytochrome into the plasma membrane. The gene for gp91-*phox* (*CYBB*) is located on chromosome Xp21.1. Two restriction fragment length polymorphisms occur in the gp91-*phox* gene, with several lesions of the gene implicated in the cause of the disease. Small deletions, nonsense or missense mutations, regulatory region mutations and splice site defects have all been identified in CGD patients.

Approximately one-third of CGD patients have defects in the p47-*phox* polypeptide. The gene for p47-*phox* (*NCF1*) is located on chromosome 7 (7q11.23) and hence inheritance is autosomal. A common defect of this gene is a dinucleotide deletion at a GTGT tandem repeat. Some patients have been shown to be homozygous for this defect, i.e. both alleles are affected.

The rest of the patients show either a defect in the p22-*phox* gene (*CYTA*) or a defect in the p67-*phox* gene (*NCF2*) with approximately 5% of each. The p22-*phox* gene has been located on chromosome 16 (16p24), with CGD being due to deletions, missense mutations and frameshifts as seen with the gp91-*phox* gene. The p67-*phox* gene (*NCF2*) has been mapped to chromosome 1 (1q25). Interestingly the polypeptide p40-*phox* is also reduced in patients lacking p67-*phox*, although no specific defects of this gene have been reported to be responsible for the occurrence of CGD.

NADPH Oxidase in Non-phagocytes and CGD

In recent years it has become apparent that NADPH oxidase activity is not confined to phagocytic cells, but has been found in cells such as fibroblasts, chondrocytes and mesangial cells as well. Therefore, there appears to be a conundrum. If NADPH oxidase has a crucial function in these cells, such as signalling, then how do patients with CGD, who have complete lack of activity of NADPH oxidase, overcome this difficulty or are the signals themselves surplus to crucial functioning? Clues first came in 1993 when Meier and her group looked at the NADPH oxidase of fibroblasts from CGD patients. It was found that although the neutrophils from an X-linked patient had severely depleted superoxide production when stimulated, the fibroblasts appeared to be unaffected. Further, the cells had a normal content of cytochrome b_{-245}, although antibodies directed against the human neutrophil cytochrome did not recognise a similar cytochrome in fibroblast membranes. These authors suggested that the low-potential cytochrome b of the NADPH oxidase from neutrophils and fibroblasts must be structurally and genetically distinct. Further work showed that the oxygen radical production of fibroblasts from CGD patients was normal when stimulated by cytokines and other agents. Other cells too have shown differences, particularly when the large cytochrome subunit, gp91-*phox*, was studied. It appears that NADPH oxidase might exist in various isoforms and it is conceivable that different forms have different functions, the non-phagocytic form having a role in cell signalling.

Other Sources of Superoxide

NADPH oxidase is not the only cellular source of superoxide and electron leakage from most electron transport chains can give up electrons to oxygen with the formation of oxygen radicals. The mitochondrial electron transport chain, for example, can lose electrons from complex I (NADH-ubiquinone oxidoreductase) or complex IV (cytochrome oxidase). However there are no reported cases of the production of oxygen radicals from this source being used in a constructive way. Further, this electron leakage has been implicated in the reduction in activity of these complexes, and it is thought that the damage caused may lead to a further increase in the production of oxygen radicals. Such a compounding effect is thought to lead to cellular death in some degenerative diseases, such as Parkinson's disease and Alzheimer's disease, and in the general ageing process.

A more likely source of oxygen free radicals is from the enzyme xanthine oxidase. This is a molybdenum- and iron-containing flavoprotein that catalyses the oxidation of hypoxanthine to xanthine and then to uric acid. Molecular oxygen is used as the oxidant and the by-product is H_2O_2. In fact, the addition of hypoxanthine and xanthine oxidase to cultured cells is a good experimental way of introducing H_2O_2 to the proximity of cells.

Evidence for Superoxide and H_2O_2 Acting as a Signal

How, if at all, are these reactive, destructive molecules being used as a way of signalling within cells? Studies are usually carried out by the addition of exogenous oxygen free radicals or by the addition of free-radical scavengers; such conditions promote or reduce the proliferation of cells in culture respectively. Crawford and colleagues noted that the addition of H_2O_2 or xanthine oxidase/xanthine stimulated the family of genes c-*fos* and c-*myc* and switched on DNA synthesis. More recently other groups have reported that H_2O_2 increased the expression of the genes c-*fos*, c-*jun*, *egr-1* and *JE* in other cell lines. There was also an increase in the level of AP-1 DNA-binding activity. AP-1 is a transcription factor, a complex composed of *jun* and *fos* gene products. Others found that the transcription factor NF-κB was activated by H_2O_2 and it was suggested that reactive oxygen compounds were serving as second messengers that directly or indirectly mediated the release of the IκB inhibitory subunit from NF–κB.

Similar signalling by reactive oxygen species has been noted in plants, where H_2O_2 induces cellular protection genes. Here, H_2O_2 acts as a diffusible element that switches on gene expression in adjacent cells, in this case the genes for glutathione S-transferase and glutathione peroxidase, both of which code for products used in the protection of cells. Very recent studies in plants suggest that superoxide and hence H_2O_2 are produced either by a peroxidase or more likely by an enzyme related to the mammalian NADPH oxidase.

As well as the activation of these transcription factors, H_2O_2 has been shown to activate MAPK. These kinases are rapidly activated in response to several external factors, promoting growth and differentiation. In neutrophils H_2O_2 caused an increase in tyrosine phosphorylation and it was concluded that activation of the MAPK was due to stimulation of tyrosine and maybe threonine phosphorylation of the kinase mediated by a MAPK activator, MEK; this was also concomitant with the inhibition by H_2O_2 of a MAP kinase phosphatase, shutting down the dephosphorylation of MAP kinase.

Reports in the literature also suggest that H_2O_2 causes the activation of both guanylyl cyclase and PLD, the latter possibly instigating the production of lipid-derived signals. Certainly, the production of prostaglandins and thromboxanes are stimulated by the presence of H_2O_2 in some cell types.

Carbon Monoxide

Recent evidence shows that guanylyl cyclase is not only under the control of NO but its activation might also be modulated by another gas, CO. For example, evidence has been presented to show that this system may be involved in the regulation of insulin secretion as well as in the control of corticotropin-releasing hormone release in the hypothalamus. Experiments with an isolated form of soluble guanylyl cyclase from bovine lung showed that CO could bind to the haem group of the enzyme to form a six coordinate complex.

Summary

A group of small inorganic molecules have been recently discovered that are produced by cells and diffuse to neighbouring cells, where they have a role in the control of many cellular functions. These molecules include NO, superoxide ions, H_2O_2 and CO.

NO is produced by the enzyme NOS, with arginine as the substrate. Cloning of the enzyme has shown that it contains areas of homology with the enzyme cytochrome P450 reductase. Both enzymes contain domains for binding NADPH, FMN and FAD and both also contain haem. The original role of NO was reported to be as a cellular relaxation factor, but it is now known that one of its functions is control of cGMP production by guanylyl cyclase.

Oxygen free radicals such as superoxide can be produced by the electron leakage of many electron transport chains, including the mitochondrial complexes. However, large amounts of superoxide are produced by the enzyme NADPH oxidase. The role of superoxide in the killing of invading pathogens is highlighted by its absence in CGD. Oxygen free radicals, including the non-radical H_2O_2, control cellular proliferation and the rates of transcription of cells, either through a direct action on transcription factors or by stimulation of MAPK type pathways.

Further Reading

Nitric Oxide

Bredt, D. S. and Snyder, S. H. (1990) Isolation of nitric oxide synthase, a calmodulin-requiring enzyme. *Proceedings of the National Academy of Sciences USA*, **87**, 682–5.

Bredt, D. S. and Snyder S. H. (1994) Nitric oxide: a physiologic messenger molecule. *Annual Review of Biochemistry*, **63**, 175–95.

Bredt, D. S., Ferris, C. D. and Snyder, S. H. (1992) Nitric oxide synthase regulatory sites: phosphorylation by cyclic AMP-dependent protein kinase, protein kinase C and calcium/calmodulin protein kinase; identification of flavin and calmodulin binding sites. *Journal of Biological Chemistry*, **267**, 10976–81.

Brune, B. and Lapetina, E. G. (1991) Phosphorylation of nitric oxide synthase by protein kinase A. *Biochemical and Biophysical Research Communications*, **181**, 921–6.

Bult, H., Boeckxstaens, G. E., Pelkmans, P. A., Jordaens, F. H., Van Maercke, Y. M. and Herman A. G. (1990) Nitric oxide as an inhibitory non-adrenergic non-cholinergic neurotransmitter. *Nature*, **345**, 346–7.

Burnett, A. L., Lowenstein, C. J., Bredt, D. S., Chang, T. S. K. and Snyder, S. H. (1992) Nitric oxide: a physiologic mediator of penile erection. *Science*, **257**, 401–3.

Cho, H. J., Xie, Q.-W., Calacay, J. *et al.* (1992) Calmodulin is a subunit of nitric oxide synthase from macrophages. *Journal of Experimental Medicine*, **176**, 599–604.

Desai, K. M., Sessa, W. C. and Vane, J. R. (1991) Involvement of nitric oxide in the relaxation of the stomach to accommodate food and fluid. *Nature*, **351**, 477–9.

Dwyer, M. A., Bredt, D. S. and Snyder, S. H. (1991) Nitric oxide synthase: irreversible inhibition by L–N^{ω}-nitroarginine in brain *in vitro* and *in vivo*. *Biochemical and Biophysical Research Communications*, **176**, 1136–41.

Evans, T., Carpenter, A. and Cohen, J. (1992) Purification of a distinctive form of endotoxin induced nitric oxide synthase from rat-liver. *Proceedings of the National Academy of Sciences USA*, **89**, 5361–5.

Furchgatt, R. F. (1995) Special topic: nitric oxide. *Annual Review of Physiology*, **57**, 659–682. A collection of several excellent articles on the production and role of nitric oxide.

Geller, D. A., Lowenstein C. J., Shapiro R. A., *et al.* (1993) Molecular cloning and expression of inducible nitric oxide synthase from human hepatocytes. *Proceedings of the National Academy of Sciences. USA*, **90**, 3491–5.

Giovanelli, J., Campos, K. L. and Kaufman, S. (1991) Tetrahydrobiopterin, a cofactor for rat cerebellar nitric oxide synthase, does not function as a reductant in the oxygenation of arginine. *Proceedings of the National Academy of Sciences USA*, **88**, 7091–5.

Hevel, J. M. and Marletta, M. A. (1992) Macrophage nitric oxide synthase: relationship between enzyme bound tetrahydrobiopterin and synthase activity. *Biochemistry*, **31**, 7160–5.

Hevel, J. M., White K. A. and Marletta, M. A. (1991) Purification of the inducible murine nitric oxide synthase: identification as a flavoprotein. *Journal of Biological Chemistry*, **266**, 22789–91.

Karupiah, G., Xie, Q.-W., Buller, R. M. L., Nathan, C., Duarte, C. and MacMicking, J. D. (1993) Inhibition of viral replication by interferon-γ induced nitric oxide synthase. *Science*, **261**, 1445–8.

Klatt, P., Schmidt, K., Uray, G. and Mayer, B. (1993) Biochemical characterisation, cofactor requirement and the role of N^{ω} hydroxy-L-arginine as an intermediate. *Journal of Biological Chemistry*, **268**, 14781–7.

Kwon, N. S., Stuehr, D. J., and Nathan C. F. (1991) Inhibition of tumour cell ribonucleotide reductase by macrophage derived nitric oxide. *Journal of Experimental Medicine*, **174**, 761–8.

McMillan, K. and Masters, B. S. S. (1993) Optical difference spectroscopy as a probe of rat brain nitric oxide synthase haem substrate interactions. *Biochemistry*, **32**, 9875–80.

McMillan, K. Bredt, D. S., Hirsch D. J., Snyder, S. H., Clark, J. E. and Masters, B. S. S. (1992) Cloned, expressed rat cerebellar nitric oxide synthase contains stoichiometric amounts of heme, which binds carbon monoxide. *Proceedings of the National Academy of Sciences USA*, **89**, 11141–5.

Mayer, B., John, M., Heinzel, B., *et al.* (1991) Brain nitric oxide synthase is a biopterin containing and flavin containing multifunctional oxidoreductase. *FEBS Letters*, **288**, 187–91.

Michel, T., Li, G. K. and Busconi, L. (1993) Phosphorylaton and subcellular translocation of endothelial nitric oxide synthase. *Proceedings of the National Academy of Sciences USA*, **90**, 6252–6.

Moncada, S. and Higgs, E. A. (eds) (1990) *Nitric Oxide from L-Arginine: A Bioregulatory System*. Elsevier, Amsterdam.

Nakane, M., Mitchell, J. A., Forstermann, U. and Murad, F. (1991) Phosphorylation by calcium calmodulin dependent protein kinase II and protein kinase C modulates the activity of nitric oxide synthase. *Biochemical and Biophysical Research Communications*, **180**, 1396–402.

Nathan, C. (1992) Nitric oxide as a secretory product of mammalian cells. *FASEB Journal*, **6**, 3051–64.

Palmer, R. M. J., Ferrige, A. G. and Moncada, S. (1987) Nitric oxide release accounts for the biological activity of endothelium derived relaxing factor. *Nature*, **327**, 524–6.

Peunova, N. and Enlkolopov, G. (1993) Amplification of calcium-induced gene transcription by nitric oxide in neuronal cells. *Nature*, **364**, 450–3.

Pollock, J. S., Forstermann, U., Mitchell, J. A., *et al.* (1991) Purification and characterisation of particulate endothelium-derived relaxing factor synthase. *Proceedings of the National Academy of Sciences USA*, **88**, 10480–4.

Pou, S., Surichamorn, W., Bredt, D. S., Snyder, S. H. and Rosen, G. M. (1992) Generation of superoxide by purified brain nitric oxide synthase. *Journal of Biological Chemistry*, **267**, 24173–6.

Reif, D. W. and Simmons, R. D. (1990) Nitric oxide mediates iron release from ferritin. *Archives of Biochemistry and Biophysics*, **283**, 537–41.

Stuehr, D. J., Cho H. J., Kwon, N. S., Weise, M. F. and Nathan, C. F. (1991) Purification and characterisation of the cytokine induced macrophage nitric oxide synthase, an FAD containing and FMN containing flavoprotein. *Proceedings of the National Academy of Sciences, USA*, **88**, 7773–7.

Weiss, G., Goosen, B., Doppler, W., *et al.* (1993) Translational regulation via iron responsive elements by the nitric oxide synthase pathway. *EMBO Journal*, **12**, 3651–7.

White, K. A. and Marletta, M. A. (1992) Nitric oxide synthase is a cytochrome P450 type hemoprotein. *Biochemistry*, **31**, 6627–31.

Yui, Y., Hattori, R., Kosuga, K, Eizawa H., Hiki, K, and Kawai, C. (1991) Purification of nitric oxide synthase from rat macrophages. *Journal of Biological Chemistry*, **266**, 12544–7.

Superoxide and Hydrogen Peroxide

Abo, A., Boyhan, A., and West, I. (1992) Reconstitution of neutrophil NADPH oxidase activity in the cell-free system by four compounds: p67-*phox*, p47-*phox*, P21RAC1 and cytochrome b_{-245}. *Journal of Biological Chemistry*, **267**, 16767–70.

Baldridge, C. W. and Gerard, R. W. (1933) The extra respiration of phagocytosis. *American Journal of Physiology*, **103**, 235–6.

Battat, L. and Francke, U. (1989) NSI RFLP at the X-linked chronic granulomatous disease locus (CYBB). *Nucleic Acids Research*, **17**, 3619.

Casimir, C. M., Bu-Ghanim H. N., Rodaway A. R. F., Bentley, D. L., Rowe, P. and Segal, A. W. (1991) Autosomal recessive chronic granulomatous disease caused by deletion at a dinucleotide repeat. *Proceedings of the National Academy of Sciences USA*, **88**, 2753–7.

Clark, R. A., Malech H. L., Gallin J. L. *et al.* (1989) Genetic variants of chronic granulomatous disease: prevalence of deficiencies of two cytosolic components of the NADPH oxidase system. *New England Journal of Medicine*, **321**, 647–52.

Crawford, D., Amstad, P., Zbinden, I. and Cerutti, P. A. (1988) Oxidant stress induces the proto-oncogenes c-*fos* and c-*myc* in mouse epidermal cells. *Oncogene*, **3**, 27–32.

Cross, A. R., Jones, O. T. G, Harper, A. M and Segal, A. W (1981) Oxidation–reduction properties of the cytochrome *b* found in the plasma membrane fraction of human neutrophils: a possible oxidase in the respiratory burst. *Biochemical Journal*, **194**, 599–606.

Cross, A. R., Parkinson, J. F. and Jones, O. T. G (1985) Mechanism of the superoxide-producing oxidase of neutrophils: O_2 is necessary for the fast reduction of cytochrome b_{-245} by NADPH. *Biochemical Journal*, **226**, 881–4.

Devary, Y., Gottlieb, R. A., Lau, L. F. and Karin, M. (1991) Rapid and preferential activation of the c-*jun* gene during the mammalian UV response. *Molecular and Cellular Biology*, **11**, 2804–11.

Didsbury, J, Weber R. F, Bokoch G. M, Evans T and Snyderman D, (1989) Rac, a novel Ras-related family of proteins that are botulinum toxin substrates. *Journal of Biological Chemistry*, **264**, 16378–82.

Dinauer, M. C., Pierce E. A, Bruns G. A. P, Curnutte J. T. and Orkin, S. II. (1990) Human neutrophil cytochrome *b* light chain (p22-*phox*) gene structure and chromosomal location: mutations in cytochrome-negative autosomal recessive chronic granulomatous disease. *Journal of Clinical Investigation*, **86**, 1729–37.

Fialkow, L., Chan, C. K., Rotin, D., Grinstein, S and Downey, G. P. (1994) Activation of the mitogen-activated protein kinase signalling pathway in neutrophils. *Journal of Biological Chemistry*, **269**, 31234–42.

Francke U., Hsieh, C.-L., Foellmer B. E., Lomax, K. J., Malech, H. L. and Leto, T. L. (1990) Genes for two autosomal recessive forms of chronic granulomatous disease assigned to 1q25 (NCF2) and 7q11.23 (NCF1). *American Journal of Human Genetics*, **47**, 483–92.

Fukumoto, Y., Kaibuchi, K. Hori, Y., *et al.* (1990) Molecular cloning and characterisation of a novel type of regulatory protein (GPI) for rho proteins, ras p21-like small GTP-binding proteins. *Oncogene*, **5**, 1321–8.

Henderson, L. M., Chappell, J. B. and Jones, O. T. G. (1988) Internal pH changes associated with the activity of NADPH oxidase of human neutrophils: further evidence for the presence of a H^+ conducting channel. *Biochemical Journal*, **251**, 563–7.

Hopkins, P. J., Bemiller, L. S. and Curnutte, J. T. (1992) Chronic granulomatous disease: diagnosis and classification at the molecular level. *Clinics in Laboratory Medicine*, **12**, 277–304.

Hurst, J. K., Loehr, T. M., Curnutte, J. T. and Rosen, H. (1991) Resonance Raman and electron paramagnetic resonance structural investigation of cytochrome b_{558}. *Journal of Biological Chemistry*, **266**, 1627–34.

Iyer, G., Islam, M. F. and Quastel, J. H. (1961) Biochemical aspects of phagocytosis. *Nature*, **192**, 535–42.

Jones, O. T. G., Jones, S. A., Hancock, J. T. and Topley, N. (1993) Composition and organisation of the NADPH oxidase of phagocytes and other cells. *Biochemical Society Transactions*, **21**, 343–6.

Jones, S. A., Jones, O. T. G., Hancock, J. T., Neubauer, A. and Topley, N. (1995) Expression of NADPH oxidase components in human glomerular mesangial cells. *Journal of the American Society of Nephrology*, **5**, 1483–91.

Kenney, R. T., Malech, H. L., Epstein, N. D., Roberts, R. L. and Leto, T. L. (1993) Characterisation of the p67-*phox* gene: genomic organisation and restriction fragment length polymorphism analysis for prenatal diagnosis in chronic granulomatous disease. *Blood*, **82**, 3739–44.

Leto, T. L., Lomax, K. J., Volpp, B. D. *et al.* (1990) Cloning of the 67kDa neutrophil oxidase factor with similarity to a non-catalytic region of P60C-SRC. *Science*, **248**, 727–30.

Levine, A., Tenhaken, R., Dixon, R. and Lamb, C. (1994) H_2O_2 from the oxidative burst orchestrates the plant hypersensitive disease resistance response. *Cell*, **79**, 583–93.

Meier, B., Cross, A. R., Hancock, J. T., Kaup, F. J. and Jones, O. T. G. (1991) Identification of a superoxide generating NADPH oxidase in human fibroblasts. *Biochemical Journal*, **275**, 241–5.

Nauseef, W. M. (1993) Cytosolic oxidase factors in the NADPH dependent oxidase of human neutrophils. *European Journal of Haematology*, **51**, 301–8.

Nose, K., Shibanuma, M., Kikuchi, K., Kageyama, H., Sakiyama, S. and Kuroki, T. (1991) Transcriptional activation of early-response genes by hydrogen peroxide in a mouse osteoblastic cell line. *European Journal of Biochemistry*, **201**, 99–106.

Nugent, J. H. A., Gratzer, W. and Segal, A. W. (1989) Identification of the haem binding subunit of cytochrome b_{-245}. *Biochemical Journal*, **264**, 921–4.

Pallister, C. J. and Hancock, J. T. (1995) Phagocyte NADPH oxidase and its role in chronic granulomatous disease. *British Journal of Biomedical Sciences*, **52**, 149–56.

Parkos, C. A., Dinauer, M. C., Walker, L. E., Allen, R.A., Jesaitis, A. J. and Orkin, S. H. (1988) Primary structure and unique expression of the 22kDa light chain of human neutrophil cytochrome b. *Proceedings of the National Academy of Sciences USA*, **85**, 3319–23.

Parkos, C. A., Dinauer, M. C., Jesaitis, A. J., Orkin, S. H. and Curnutte, J. T. (1989) Absence of both the 91kDa and 22kDa subunits of human neutrophil cytochrome b in two genetic forms of chronic granulomatous disease. *Blood*, **73**, 1416–20.

Radeke H. H., Cross, A. R., Hancock, J. T., *et al.* (1991) Functional expression of NADPH oxidase components, α and β subunits of cytochrome b_{-245}, 45kDa flavoprotein, by intrinsic glomerular mesangial cells. *Journal of Biological Chemistry*, **31**, 21025–9.

Rotrosen, D., Yeung, C. L., Leto, T. L., Malech, H. L. and Kwong, C. H. (1992) Cytochrome b_{558}, the flavin binding component of the phagocyte NADPH oxidase is a flavoprotein. *Science*, **256**, 1459–62.

Royer-Pokora, B., Kunkel, L. M., Monaco, A. P., *et al.* (1986) Cloning the gene for the inherited human disorder chronic granulomatous disease on the basis of its chromosomal location. *Nature*, **322**, 32–8

Sbarra, A. J. and Karnovsky, M. L. (1959) The molecular basis of phagocytosis. *Journal of Biological Chemistry*, **234**, 1355–62.

Schreck, R., Rieber, P. and Baeuerle, P. A. (1991) Reactive oxygen intermediates as apparently widely used messengers in the activation of the NF-κB transcription factor and HIV-1. *EMBO Journal*, **10**, 2247–58.

Segal, A.W. (1987) Absence of both cytochrome b_{-245} subunits from neutrophils in X-linked chronic granulomatous disease. *Nature*, **326**, 88–91.

Segal, A. W. and Abo, A. (1993) The biochemical basis of NADPH oxidase in phagocytes. *Trends in Biochemical Sciences*, **18**, 43–7.

Segal, A. W. and Jones, O. T. G., (1978) Novel cytochrome b system in phagocytic vacuoles of human granulocytes. *Nature*, **276**, 515–7.

Segal, A. W., Cross, A. R., Garcia, R. C., *et al.* (1983) Absence of cytochrome b_{-245} in chronic granulomatous disease: a multi European evaluation of its incidence and relevance. *New England Journal of Medicine*, **308**, 245–51.

Segal, A. W., West, I., Wientjes, F. B. *et al.* (1992) Cytochrome b$_{-245}$ is a flavoprotein containing FAD and the NADPH binding site of the microbial oxidase of phagocytes. *Biochemical Journal*, **284**, 781–8.

Stevenson, M. A., Pollock, S. S., Coleman, C. N. and Calderwood, S. K. (1994) X-irradiation, phorbol esters, and H$_2$O$_2$ stimulate mitogen-activated protein kinase activity in NIH-3T3 cells through the formation of reactive oxygen intermediates. *Cancer Research*, **54**, 12–15.

Sumimoto, H., Sakamoto, N., Nozaki, M., Sakaki, Y., Takeshige, K. and Minakami, S. (1992) Cytochrome b$_{558}$, a component of the phagocyte NADPH oxidase is a flavoprotein. *Biochemical and Biophysical Research Communications*, **186**, 1368–75.

Thrasher, A. J., Keep, N. H., Wientjes, F. and Segal, A. W. (1995) Chronic granulomatous disease. *Biochimica et Biophysica Acta*, **1227**, 1–24.

Volpp, B. D. and Lin, Y. (1993) *In vitro* molecular reconstitution of the respiratory burst in B lymphoblasts from p47-*phox* deficient chronic granulomatous disease. *Journal of Clinical Investigation*, **91**, 201–7.

Volpp, B. D., Nauseef, W. M., Donelson, J. E., Moser, D. R. and Clark, R. A. (1989) Cloning the cDNA and functional expression of the 47kDa cytosolic component of human neutrophil respiratory burst oxidase. *Proceedings of the National Academy of Sciences USA*, **86**, 7195–9.

Wientjes, F. B., Hsuan, J. J., Totty, N. F. and Segal, A. W. (1993) p40-*phox*, a third cytosolic component of the activation complex of the NADPH oxidase to contain src homology 3 domains. *Biochemical Journal*, **296**, 557–61.

Wood, P.M. (1987) The two redox potentials for oxygen reduction to superoxide. *Trends in Biochemical Sciences.* **12**, 250–1.

Yamaguchi, T., Hayakawa, T., Kaneda, M., Kakinuma, K. and Yoshikawa, A. (1989) Purification and some properties of the small subunit of cytochrome b$_{558}$ from human neutrophils. *Journal of Biological Chemistry*, **264**, 112–8.

Carbon Monoxide

Pozzoli, G., Mancuso, C., Mirtella, A., Preziosi, P., Grossman, A. B. and Navarra, P. (1994) Carbon monoxide as a novel neuroendocrine modulator: inhibition of stimulated corticotropin-releasing hormone release from acute rat hypothalamic explants. *Endocrinology*, **135**, 2314–17.

Schmidt, H. H. H. W. (1992) NO-built, CO and bullet-OH, endogenous soluble guanylyl cyclase activating factors. *FEBS Letters*, **307**, 102–7.

Shragalevine, Z., Galron, R. and Sokolovsky, M. (1994) Cyclic GMP formation in rat cerebellar slices is stimulated by endothelins via nitric oxide formation and by sarafaotoxins via formation of carbon monoxide. *Biochemistry*, **33**, 14656–9.

Stone, J. R. and Marletta, M. A. (1994) Soluble guanylate cyclase from bovine lung: activation with nitric oxide and carbon monoxide and spectral characterization of the ferrous and ferric states. *Biochemistry*, **33**, 5636–40.

Vara, E., Ariasdiaz, J., Garcis, C. and Balibrea, J. L. (1994) Evidence for a cGMP-dependent, both carbon monoxide and nitric oxide mediated, signalling system in the regulation of islet insulin secretion. *Diabetologia*, **37**, A112.

9

Insulin and the Signal-transduction Cascades it Invokes

In the preceding chapters the separate components of possible signal-transduction pathways have been discussed. Here, the way in which many of these components come together is illustrated using insulin as an example of a hormone recognised by the extracellular surface of a cell and which, through a cascade of various messengers, results in many cellular effects. Some of the effects resulting from insulin binding are listed in Table 9.1. Some of the insulin effects are through modulation

Table 9.1. Some of the cellular effects induced by insulin.

Phosphorylation of IRS-1
Activation of phosphatidylinositol 3-kinase
Activation of Ras
Phosphorylation of kinases: MAP kinases, ribosomal S6 kinase
Phosphorylation of phosphatases
Phosphorylation of metabolic enzymes
Dephosphorylation of metabolic enzymes
Translocation of proteins: GLUT-4, insulin receptors
Regulation of gene expression
Modulation of protein synthesis

of cytosolic enzymes, while other effects include control of gene expression inside the nucleus. In general, insulin promotes anabolic processes of a cell while causing a reduction in catabolic processes. For example, insulin promotes the synthesis of glycogen and fatty acids while simultaneously inhibiting their breakdown. An alteration of insulin signalling, its production, its detection or subsequent signalling, can lead to the disease diabetes mellitus.

Insulin is produced by the cells of the Islets of Langerhans. These are cell clusters found in the pancreas. As well as insulin, these cells are responsible for the production of another hormone, glucagon, which is a single polypeptide chain hormone. Glucagon and insulin are responsible for the regulation of the concentration of glucose in the bloodstream of mammals.

The route for the production of insulin was studied in a human islet-cell tumour, where insulin was produced in large amounts. Using tritiated leucine incorporation into the insulin polypeptide allowed analysis of the various steps. Production starts with the synthesis of a single polypeptide chain known as preproinsulin. At the N-terminal end of this molecule is a 19-amino acid signal sequence, which is relatively hydrophobic and directs the polypeptide to the ER. In the lumen of the ER the 19 amino acids are proteolytically removed to form proinsulin. Proinsulin

Figure 9.1

Amino acids sequence of insulin. The disulphide bonds are indicated by the solid lines, while the connecting peptide that is removed is shown by the dotted line.

subsequently passes through the Golgi apparatus of the cell and into secretory granules, where a 33–amino acid stretch of polypeptide is removed from the middle of the chain. When insulin from different species is analysed the sequence removed is found to contain a Lys-Arg sequence at the C-terminal end with an Arg-Arg sequence at the N-terminal end, which must serve as recognition sequences for the protease responsible for the cleavage. Once removed, this leaves two polypeptide chains, an A chain of 21 amino acids and a B chain of 30 amino acids, which are connected by two disulphide bridges. A third disulphide bridge also spans across two cysteine residues in the smaller chain of amino acids, the A chain (Figure 9.1). The amino acid sequence of insulin was determined in 1953 by Frederick Sanger.

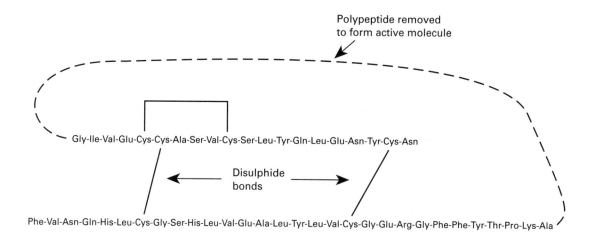

Release of insulin from the secretory granules into the bloodstream is under the control of other hormones and neuronal signals. The level of insulin measurable in the blood is usually very low, 10^{-10} mol/l.

The three-dimensional structure of insulin was determined to a resolution of 1.9 Å by Dorothy Hodgkin's laboratory. As well as the three disulphide bonds, the structure is held together by salt links and hydrogen bonds. The molecule assumes a basically globular structure, with only the N-terminal and C-terminal ends of the B chain projecting into solution.

A second molecule that shares great similarity to insulin is insulin-like growth factor 1 (IGF-1). This is a hormone similar in both sequence and structure to insulin but produced in the liver. Its release is controlled by pituitary hormones and its role seems to be in the control of an organism's growth.

As stated earlier the effects of insulin are fairly numerous. Some of the effects are seen relatively quickly, while others can take a matter of hours. The immediate effects of insulin include modulation of the activities of various enzymes in different metabolic pathways, some of which are listed in Table 9.2, and an increase in the rate of glucose uptake.

Table 9.2. Some enzymes that have their activity modulated through the action of insulin.

Enzyme	Metabolic pathway involved	Effect of insulin on activity
Phosphorylase kinase	Glycogen metabolism	Decrease
Pyruvate kinase	Glycolysis	Decrease
Lipase	Lipid breakdown	Decrease
Glycogen synthase	Glycogen synthesis	Increase
Pyruvate dehydrogenase	Citric acid cycle	Increase
Acetyl–CoA carboxylase	Fatty acid synthesis	Increase
Hydroxymethylglutaryl–CoA reductase	Cholesterol synthesis	Increase

Usually such effects are stimulated by blood insulin levels in the order of 10^{-9} to 10^{-10} mol/l. Longer exposure of cells to higher concentrations of insulin (approximately 10^{-8} mol/l) induces protein synthesis of the enzymes involved in glycogen synthesis in the liver and the enzymes needed in triglycerol synthesis in adipose tissue, and stimulates proliferation in some cells such as fibroblasts. However, such high concentrations are generally not physiological.

Insulin is detected by the target cell mainly via the insulin receptor. It was first purified by Pedro Cuatrecasa. The receptor has a binding affinity for insulin of approximately 10^{-10} mol/l, the same order of magnitude as the concentration of insulin in the blood. The receptor is a glycoprotein that contains two large α subunits of 135 kDa and two smaller β subunits of 95 kDa, and therefore is described as having an $\alpha_2\beta_2$ subunit structure. The β subunits are integral to the plasma membrane while the α subunits reside on the outside of the cell membrane and are held to the β subunits by disulphide bonds (Figure 9.2). The α and β subunits of the receptor are synthesised as a single polypeptide chain of 1382 amino acids. The precursor peptide has a signal peptide at the N-terminal end that is cleaved off leaving the α subunit, followed by a tetrapeptide with the sequence Arg-Lys-Arg-Arg, followed by the sequence of the β subunit. The tetrapeptide is highly basic and serves as a recognition signal for the protease that processes the polypeptide.

Insulin also binds to the IGF-1 receptor. This receptor is very similar to the insulin receptor but insulin binds to the IGF-1 receptor approximately 100 times less well than IGF-1. Conversely, IGF-1 binds to the insulin receptor but again about 100 times less than insulin.

One of the effects of insulin is to increase the capacity of the cell to take up glucose; this results in the lowering of blood glucose concentrations, the so-called hypoglycaemic effect. Within approximately 15 min of the application of insulin to adipocytes, their rate of uptake of glucose increases by 10–20 fold. This short time span is not sufficient to allow for the synthesis of new protein and hence new glucose transporters. However, analysis has shown that the number of glucose transporters found in the plasma membrane greatly increases during this time. Glucose transporters, otherwise known as the GLUT-4 protein, are synthesised and targeted to endosome-like vesicles. They remain as integral membrane proteins of these vesicles until the cell is stimulated by insulin. Once stimulated, the vesicles are transported to the plasma membrane, where the membranes fuse and the GLUT-4

Figure 9.2
Model of the structure
of the insulin receptor.

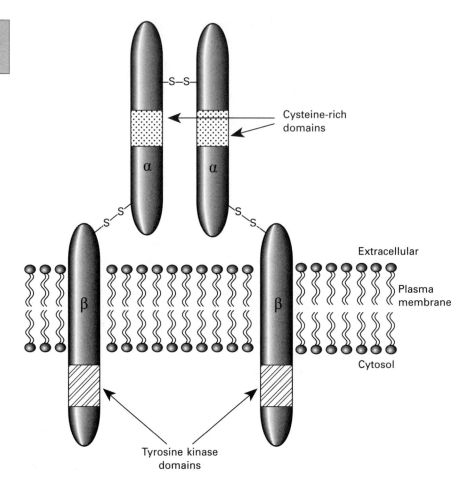

Cysteine-rich
domains

Extracellular

Plasma
membrane

Cytosol

Tyrosine kinase
domains

transporters become part of the plasma membrane, where they function in the rapid uptake of glucose. On the removal of the insulin signal, the GLUT-4 proteins undergo endocytosis into the interior of the cell, and so the uptake of glucose can once again be reduced. Recent work, where the transporters have been tagged with a fluorescent marker, has allowed this movement of the GLUT-4 polypeptides to be visualised using a confocal microscope. However, the link between the insulin receptor and translocation of the endosome-like vesicles has yet to be established, although a monomeric G protein, but not Ras, and two protein kinases have been suggested to be involved.

The signal-transduction mechanisms leading to the other effects of insulin have been much more fully elucidated. The α subunits of the receptor contain cysteine-rich domains and it is on these subunits that insulin binding takes place. The β subunits, on the other hand, contain domains with tyrosine kinase activity and hence the receptors are classed as RTKs. On binding of insulin to the receptor, the first response is that the receptor phosphorylates itself on at least three tyrosine residues, i.e. autophosphorylation (Figure 9.3). This has two results. Firstly, it activates the receptor to phosphorylate other cellular proteins. Secondly, even if the

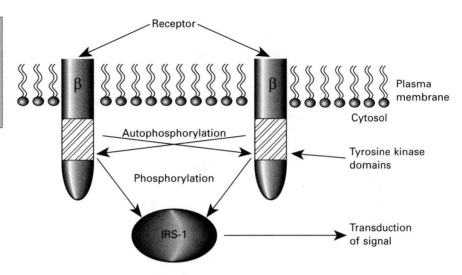

Figure 9.3
Schematic representation of the phosphorylation events that result from activation of the insulin receptor. IRS-1, insulin receptor substrate 1.

insulin is subsequently removed from the receptor, the receptor remains active unless it is dephosphorylated by a phosphatase.

One of the proposed proteins phosphorylated as a result of activation of the insulin receptor is a protein phosphatase. One such phosphatase is PP1G, which is found associated with glycogen particles. The activity of PP1G is increased if it is phosphorylated on a serine residue. Therefore, as the insulin receptor is a tyrosine kinase it is proposed that another kinase is involved in the signal cascade, one such kinase being referred to as insulin-stimulated protein kinase (ISPK). Binding of insulin to its receptors results in the dephosphorylation and therefore activation of glycogen synthase, and it is therefore proposed that the activation of a phosphatase such as PP1G through a cascade induced by the insulin receptor is the key to this mechanism.

However, one of the key events in signal transduction from the insulin receptor is the multiphosphorylation of a protein known as IRS-1 (see Figure 9.3). This is a protein of 130 kDa and can be phosphorylated by either the insulin receptor or the IGF-1 receptor. IRS-1 is highly serine phosphorylated and also can be highly tyrosine phosphorylated by these receptors. Six of the sites of tyrosine phosphorylation on this protein have the neighbouring sequence of Tyr-(Pro)-Met-X-Met, where X is a variety of amino acids. Once phosphorylated and therefore activated, IRS-1 interacts with, and modulates the activity of, different pathways. IRS-1 binds to the SH2 domain of the regulatory subunit of PtdIns 3-kinase, causing its activation. PtdIns 3-kinase is also phosphorylated on a tyrosine by the insulin receptor. As discussed in Chapter 6, this leads to the production in the membrane of PtdIns lipids which are phosphorylated on the 3 position of the inositol ring. The exact mechanism then invoked is not known, but several actions of insulin on cells have been shown to be inhibited by the fungal metabolite wortmannin, which is a reasonably specific inhibitor of PtdIns 3-kinase.

Better understood are other events resulting from the activation of IRS-1. IRS-1 acts as a relay protein, being phosphorylated by the insulin receptor but not actually binding to it. Once phosphorylated it then interacts with further proteins. As it is phosphorylated on tyrosine residues this creates binding sites for SH2 domains (see

189

Chapter 1). Through its SH2 domains, an adaptor protein called GRB2 binds to the phosphorylated IRS-1. The GRB2 protein also contains two SH3 domains, which bind to a GNRP, in this case Sos. Sos, once activated, catalyses the release of GDP from the monomeric G protein Ras and therefore allows the G protein to take up its active state in which GTP is bound. Ras then activates the kinase Raf, which leads to the phosphorylation and activation of the MAPK cascade. This scheme is illustrated in Figure 9.4.

Figure 9.4
One of the signalling cascades associated with insulin detection. Binding of insulin to its receptor leads to autophosphorylation of the receptor and phosphorylation of a relay protein, insulin receptor substrate 1 (IRS-1). The phosphotyrosines formed on IRS-1 interact with the SH2 domains of the adaptor GRB2 and, like the epidermal growth factor receptor pathway, leads to sequential involvement of Sos, Ras and then a mitogen–activated protein (MAP) kinase cascade that transmits the signal into the cell, often resulting in enhanced transcription. Activation of phosphatidylinsitol 3–kinase also occurs but the significance of this is unclear.

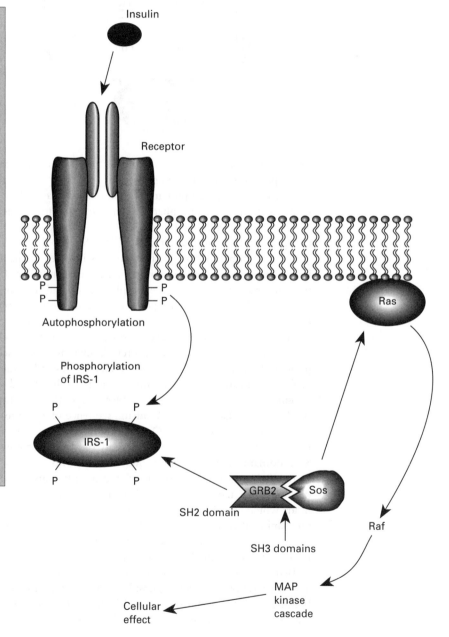

One of the end-results of activation of this cascade is the alteration of gene expression. The promoter regions of the genes controlled by insulin have also come under close scrutiny. These insulin-responsive elements (IREs) have been determined in several genes, including those for phosphoenolpyruvate carboxylase, amylase, liver pyruvate kinase and glyceraldehyde 3-phosphate dehydrogenase. In general they seem to contain two 8-base pair elements that are AT-rich, both of which need to be present, although some genes, such as that for liver pyruvate kinase, do not follow this pattern. By using fusion of the luciferase gene to the relevant promoters, looking for light emission from cells where the gene product is formed and then by transforming cells with mutated forms of the proteins involved in the MAPK pathway, Tavaré and colleagues have illustrated the importance of some of the above signal-transduction components in insulin signalling.

Once activated by the insulin receptor, IRS-1 also interacts with a protein called Syp. This is a tyrosine phosphatase and it has been proposed that it is involved in the dephosphorylation of the IRS-1 protein itself and therefore involved in turning off the insulin-induced signal.

The insulin receptor is also internalised and downregulated. ^{125}I-labelled insulin is taken into the cell quite rapidly. After 5 min up to 30% of the insulin bound to the cell is associated with internal vesicles. Internalised insulin becomes dissociated from the receptor and is generally destroyed by the cell. While still in the phosphorylated state the receptor continues to signal to the cell, even though it has been internalised, although normally it becomes dephosphorylated and may be translocated back to the plasma membrane for another round of signalling. Several phosphatases have been identified that may be involved in the dephosphorylation of the insulin receptor.

Summary

Insulin has varied and profound effects on many aspects of a cell's function, including the control of metabolic pathways, the stimulation of protein translocation and the control of gene expression. Insulin is composed of two peptide chains, held together by disulphide bonds, hydrogen bonds and salt bridges. Cells detect insulin mainly via the insulin receptor, but insulin also binds to the receptor for IGF-1.

The insulin receptor is a tetramer with an $\alpha_2\beta_2$ subunit structure. The α subunits are on the outside of the cell and are responsible for insulin binding, while the β subunits contain transmembrane domains along with tyrosine kinase domains. Binding of insulin to its receptor results in autophosphorylation of the receptor on tyrosine residues and also in the phosphorylation of other cellular proteins, such as IRS-1. Activation of IRS1 leads to activation of other proteins through the interaction of its phosphotyrosyl residues with the SH2 domains of other polypeptides. These proteins include PtdIns 3-kinase, leading to an as yet unexplained signalling pathway. IRS-1 also interacts with an adaptor protein such as GRB2, which leads to activation of MAPK cascades and the modulation of gene expression.

The insulin signal is turned off by the internalisation of the receptor, which may be recycled back to the plasma membrane. Integral to this process is the dephosphorylation of various proteins involved in insulin signalling.

Further Reading

Dent, P., Lavoinne, A., Nakielny, S., Caudwell, F. B., Watt, P. and Cohen, P. (1990) The molecular mechanism by which insulin stimulates glycogen synthesis in mammalian skeletal muscle. *Nature*, **348**, 302–8.

Hashimoto, N., Feener, E. P., Zhang, W.-R. and Goldstein, B. J. (1992) Insulin receptor protein-tyrosine phosphatases. *Journal of Biological Chemistry*, **267**, 13811–14.

Kimball, S. R., Vary, T. C. and Jefferson, L. S. (1994) Regulation of protein synthesis by insulin. *Annual Review of Physiology*, **56**, 321–48.

Lawrence J. C. Jr (1992) Signal transduction and protein phosphorylation in the regulation of cellular metabolism by insulin. *Annual Review of Physiology*, **54**, 177–93.

Makino, H., Manganiello, V. C. and Kono, T. (1994) Roles of ATP in insulin actions. *Annual Review of Physiology*, **56**, 273–95.

Myers, M. G. and White, M. F. (1996) Insulin signal transduction and the IRS proteins. *Annual Review of Pharmacology and Toxicology*, **36**, 615–58.

Roth, R. A. (1990) Insulin receptor structure. *Handbook of Experimental Pharmacology*, **92**, 169-81.

Rothenberg, P., White, M. F. and Kahn, C. R. (1990) The insulin receptor tyrosine kinase. *Handbook of Experimental Pharmacology*, **92**, 209–36.

Rutter, G. A., White, M. R. H. and Tavaré, J. M. (1995) Involvement of MAP kinase in insulin signalling revealed by noninvasive imaging of luciferase gene-expression in single living cells. *Current Biology*, **5**, 890–9.

Sun, X. J., Rothenberg, P., Kahn, C. R., *et al.* (1991) Structure of the insulin receptor substrate IRS-1 defines a unique signal transduction protein. *Nature*, **352**, 73–7.

10

Perception of our Environment: Photoreception and the Detection of Chemicals

Introduction

Like individual cells, whole organisms need to detect the properties of their environment. This might be the detection of a source of food, the sensing of something that may be harmful, such as a predator, or the detection of a pheromone as discussed in Chapter 2. Many organisms have adapted their awareness of their environment to suit their niche in nature, for example some animals such as moles and bats have very poor eyesight, not easily detecting and perceiving light, because they either live underground or hunt at night where little light is available, while others might have an acute sense of smell or very efficient hearing, such as animals that graze in the open and need to sense the arrival of a predator before it has a chance to get too close.

The signalling involved in such perception systems is often akin to those used by individual cells within the body, with the use of specific receptors tuned to the particular environmental factor to be detected, and the transmission of a change in the receptor into the interior of the cell and the subsequent cellular response. Analogous systems are used whether it is a bacterium sensing the presence of a chemical, leading to chemotaxis, or the eye of a human. A look at some of these systems of perception illustrates how the interaction and coordinated use of several signalling pathways leads to a response.

Photoreception

A sense many of us take for granted is the ability to see, but the perception of light is not unique to mammals. Lower animals and plants also can and need to respond to light in order to thrive.

Photosensitivity of Plants

Plants need to perceive light and respond to it in order to survive. Leaves on a tree, for example, are optimally aligned to maximise the amount of light captured for photosynthesis. If a leaf is in the shade, the plant might need to increase the leaf area exposed to direct sunlight. Similarly, plants are able to detect the time of day, for example flowers opening and closing, and also the time of year.

Plants have four main ways of monitoring light: phytochromes, UV-B receptors, UV-A receptors and blue light receptors.

Phytochromes of higher plants are coded for by at least five genes, assigned the names *PHYA–PHYE*. Phytochromes are responsible for detecting red and far-red light. However, the exact biochemical signalling that results from their stimulation is unknown. Evidence has been put forward that suggests that phytochromes act as kinases or at least contain kinase activity but this has never been substantiated, although sequence alignment studies and the alteration of specific amino acids in the polypeptides is now giving an insight into which regions of the molecules may be important for their functioning. The trimeric G protein family leading to increases in Ca^{2+}, associated with calmodulin and increases in cGMP, have been suggested to be involved in the transduction of the signal in the cell. cGMP and Ca^{2+} probably have their main effect through their influence on the levels of gene expression.

A protein designated as HY4 has been cloned from *A. thaliana* and is likely to be the blue-light photoreceptor, although its exact mode of action and the intracellular signalling involved have yet to be elucidated.

Photodetection in the Eye

Photoreception in humans is the responsibility of two types of specialised cells, the rods and cones, so named because of their shape. These cells form a layer called the retina at the back of the eye on to which the light is focused by the lens. The rods are used in the perception of a wide range of wavelengths of light at low light levels whereas the cones function in brighter light and fall into three classes, each with their own wavelength sensitivity. One class senses blue light, one green light and a third red light. In humans the wavelength maxima for the three classes of cones are 426, 530 and 560 nm, although other species such as fish have slightly different wavelength maxima: 455, 530 and 625 nm.

The cells themselves are bipolar, having an outer segment and an inner segment. The outer segment is basically a bag containing flattened membranous sacks called discs, which lie one on top of the other perpendicular to the plane of the incident light. A rod may contain up to 1000 of these discs, which are only about 16 nm thick. The inner segment contains the main body of the cell including the machinery of the ER, mitochondria and a nucleus, but more importantly a synaptic body used for the transmission of any created signal (Figure 10.1).

The photoreceptive molecule (chromophore) in the rods is rhodopsin. This protein, called opsin, contains 11-*cis*-retinal as a prosthetic group (Figure 10.2). 11-*cis*-Retinal is derived from vitamin A (all-*trans*-retinol) obtained in the diet. In

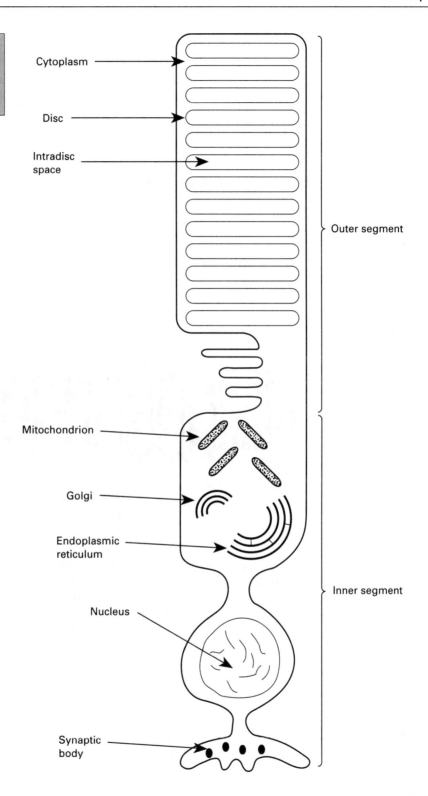

Figure 10.1
A rod cell, one of the cells inolved in photoreception in mammals.

Cytoplasm

Disc

Intradisc space

Outer segment

Mitochondrion

Golgi

Endoplasmic reticulum

Inner segment

Nucleus

Synaptic body

Figure 10.2
Molecular structure of
11-*cis*-retinal.

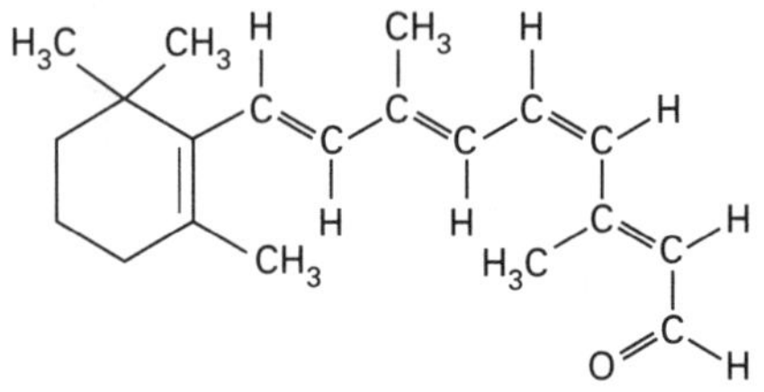

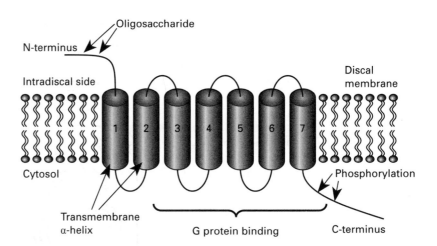

the protein it is attached via a protonated Schiff base linkage to a lysine residue and it is this interaction, along with others, that gives 11-*cis*-retinal an absorbance maximum of approximately 500 nm. Interestingly, rhodopsin also has an extremely large extinction coefficient, which means that it is very efficient at absorbing light.

The opsin protein is like many other membrane proteins in having seven membrane-spanning α-helices (Figure 10.3). The retinal lies in the centre of the α-

Figure 10.3
Domain structure of
rhodopsin showing the
similarity to seven
transmembrane-
spanning receptors.

helical section and lies in the plane of the membrane. The N-terminal end of the polypeptide, which contains two N-linked oligosaccharides, lies in the intradiscal space, while the C-terminal end lies on the cytosolic side of the membrane. It is this region that is important in interactions with other proteins that transmit the signal to the rest of the cells and ultimately to the nervous system, for example it is this region where the G protein binds and also where phosphorylation on serine and threonine residues occurs, leading to deactivation of the molecule.

On receiving a photon of light 11-*cis*-retinal photoisomerises to all-*trans*-retinal, inducing a conformational change in the molecule. This means that the Schiff base linkage to the protein moves by approximately 5 Å. This new molecule, called bathorhodopsin, undergoes several conformational changes, which can be monitored by changes in the maximal absorbance wavelength, until metarhodopsin II, otherwise known as photoexcited rhodopsin, is formed in which the Schiff base linkage has become unprotonated. It is the creation of this form that triggers the resultant signalling cascade.

Rhodopsin is reformed by the hydrolysis of the all-*trans*-retinal, its reconversion to 11-*cis*-retinal, which involves all-*trans*-retinol as an intermediate, and then relinkage to opsin.

The same light-sensitive system is also used in the cones, except here the opsin-like molecule has three hydroxyl group-containing residues that surround the 11-*cis*-retinal and the alteration of its local environment is enough to alter its light-absorbing characteristics and its wavelength maximum.

The C-terminal end of the rhodopsin molecule in the unexcited state is associated with a member of the trimeric G protein family, transducin or G_t. This was in fact only the second G protein to be discovered, by both Mark Bitensky and Lubert Stryer. It is composed of an α subunit of 39 kDa, a β subunit of 37 kDa and a γ subunit of 8.5 kDa. The excited rhodopsin acts effectively like a GTP exchange factor, exchanging the GDP bound to the G protein for a GTP so causing its activation. The G protein dissociates into $G_{t\alpha}$ and $G_{t\beta\gamma}$ subunits. The $G_{t\alpha}$ subunit interacts with the inhibitory peptide of PDE and so reduces the inhibited state of this enzyme. This enzyme normally resides as a complex of two catalytic subunits (α and β) that are inhibited by the presence of two inhibitory peptides, the γ subunits. This inhibitory restraint is removed by the interaction with $G_{t\alpha}$. The activated PDE very rapidly and efficiently hydrolyses cGMP in the cytosol of the cell's outer segment (Figure 10.4).

The subsequent drop in the cytosolic concentration of cGMP leads to closure of cation-specific channels in the plasma membrane of the cells. In the dark these channels are kept open by the binding of cGMP. The channel is a multipolypeptide complex, each subunit having a molecular mass of approximately 80 kDa. The cGMP binding is extremely cooperative, opening of the channel requiring interaction with at least three cGMP molecules. The open channels serve as a route for the return of Na^+ ions into the outer segment of the cells down a large electrochemical gradient. Na^+/K^+-ATPases in the plasma membrane of the inner segments of the cells are used to maintain this gradient. Closure of the channels via the drop in cGMP levels means that the Na^+ ions are no longer able to return into the outer segment and this results in hyperpolarisation of the plasma membrane. This hyperpolarisation can be as great as 1 mV and is sensed by the synaptic body at the base of the cell and transmitted to the neurones, and so to the brain.

Although the $G_{t\alpha}$ subunit is the active part of the G protein, the $G_{t\beta\gamma}$ subunit also interacts with a protein, known as the 33K protein, or phosducin. Despite its name, this protein actually has a molecular mass of approximately 28 kDa. It is phosphorylated by cAPK on a serine residue and dephosphorylated by phosphatase 2A; interestingly, this phosphorylation is light dependent, the protein being most highly phosphorylated in the dark. It is thought that phosducin may regulate the recycling of the $G_{t\alpha}$ subunit and so modulate the amount of G protein available for interaction with rhodopsin, this function itself being regulated by a phosphorylation event.

Massive amplification of signalling is seen in this system. One rhodopsin molecule, once activated, can stimulate up to 500 transducin G proteins, while massive amplification is also seen in the cleavage of cGMP by the extremely active phosphodiesterase.

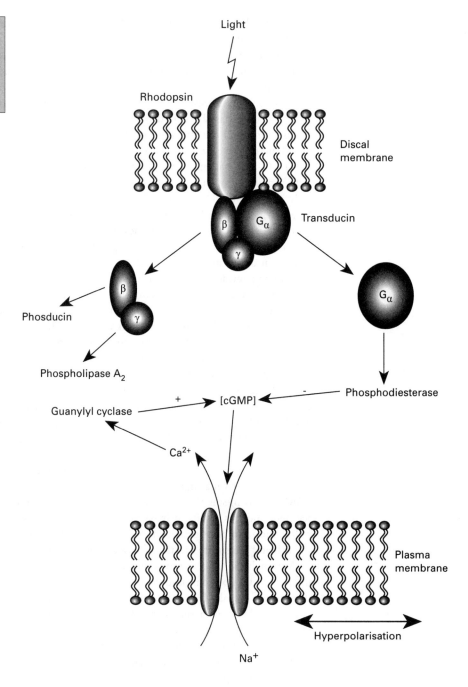

Figure 10.4
Signal-transduction pathway in rod cells showing how light leads to hyperpolarisation of the plasma membrane.

The system is turned off by several simultaneous events. Ca^{2+} is normally pumped out of the cell via an exchanger but its re-entry is normally allowed into the outer segments via the cation-specific channels. On closure of these channels following the drop of cGMP, intracellular Ca^{2+} levels drop and this in turn stimulates the synthesis of cGMP by the activation of guanylyl cyclase. Resurrection of the cGMP levels will once again open the cation channels and end the hyperpolarisation of the membranes.

Transducin has, like all heterotrimeric G proteins, endogenous GTPase activity and thus any bound GTP is hydrolysed, allowing the re-formation of the trimeric inactive G protein, assuming phosducin has not had an influence on $G_{\beta\gamma}$ levels. Transducin could then be reactivated by rhodopsin, if it was not turned off, but in fact rhodopsin is phosphorylated by rhodopsin kinase on several serine and threonine residues on the C-terminal end of the polypeptide. Once phosphorylated, rhodopsin interacts with a polypeptide called arrestin. Interaction with arrestin prevents rhodopsin from interacting with transducin and so prevents further activation of the cascade. It is interesting to note that a similar mechanism is seen with β-adrenergic receptors; in fact βARK phosphorylates the rhopdopsin molecule in a light-dependent manner, highlighting the structural similarity between these receptor molecules.

Although the signalling mechanisms discussed have been mainly elucidated by studies on the rods from the eye, the cones have an analogous system. However, subtle differences are apparent, such as in the G proteins used. Rods express the G protein subunits α_{t-r}, β_1 and γ_1, while cones express a G protein consisting of a cone-specific α_t, β_3, and unknown γ subunit.

To complicate matters even more, PLA_2 activity in rod outer segments is stimulated by light. It has been found that transducin $\beta\gamma$ subunits increase PLA_2 activity, which is inhibited by transducin α subunits.

It can be seen therefore that light perception in the eye serves as a good example of how several signalling components work together to result in a coordinated response (see Figure 10.4). Here, a light receptor, a G protein, an enzyme, a second messenger and an ion channel are all used to transmit the signal, with the cytosolic Ca^{2+} concentration and another enzyme being involved in the return to the ground state.

It must not be forgotten that other systems and components have also been suggested to be involved, including phosducin and PLA_2. Work studying the photoreceptors of amphibians has shown that the activity of NOS and the production of NO has an influence on the metabolism of cGMP and therefore may well be involved in the adaptive nature of the eye. It is not just the absolute intensity of light to which we are sensitive but rather changes in intensity and therefore a biochemical mechanism needs to explain how our eyes can so readily adapt to wide ranges of light intensity, maybe as much as a 10 000-fold difference.

The signalling pathways involved in the light-sensing organs of lower animals differ somewhat from those of vertebrates. In the fruit fly *Drosophila melanogaster* rhodopsin activates a trimeric G protein of the G_q type, which in turn activates PLC. The PLC catalyses the breakdown of $PtdInsP_2$ in the membrane and therefore produces DAG and $InsP_3$. The $InsP_3$ opens Ca^{2+} channels and allows the release of Ca^{2+} from internal stores. However, the exact way in which the Ca^{2+} causes the propagation of the signal remains unclear. Photoadaptation and recovery also need the participation of an eye-specific PKC.

Detection of Chemicals

Prokaryotes

Bacteria, such as *E. coli*, need to detect molecules in the extracellular media and respond to their presence. Such molecules include attractants such as the sugars galactose and ribose, the amino acids serine, alanine and glycine as well as dipeptides. Repellent molecules include acetate, leucine, indole and metals such as nickel and cobalt. The presence of these molecules is detected by receptors known as methyl-accepting chemotaxis proteins. These proteins form four homologous groups, known as Tsr, Tar, Trg and Tap, and all span the inner bacterial membrane. Sequence homology between the members of these receptor groups shows that the N-terminal end contains the receptor domain that interacts with the ligand, with a membrane-spanning domain on each side. The C-terminal end contains the domain that propagates the signal into the cell. The C-terminal end is also the site of methylation, which seems to be responsible for the sensitivity of receptor, enabling it to be receptive to a concentration gradient of the ligand rather than simply responding to an on–off signal. Methylation is catalysed by a methyltransferase known as CheR, while demethylation is catalysed by CheB-phosphate (CheB-P). The methyl group is supplied from *S*-adenosylmethionine while the targets for transfer are glutamate residues. Different receptor classes are methylated differently, but, in general, between four and six methylation sites are seen. CheB-P, a methylesterase, removes the methyl group forming methanol. The CheB protein is phosphorylated in its most active state, i.e. as CheB-P, the non-phosphorylated polypeptide being a target of a kinase, CheA.

The receptors exist in the membrane as dimers, although binding of one subunit, with no major conformational changes in structure, is sufficient to cause an effect. The receptors also appear to be arranged in clusters on the cell, being particularly aggregated at the cell pole. However, this arrangement does not seem logical and has not been explained.

The bacterium responds to the presence of the ligand by rotating its flagellum in a counter-clockwise direction, causing a smooth swimming action and directional movement of the cell. In the absence of the ligand, the flagellum rotates clockwise and a tumbling movement is seen, leading to the apparent wandering of the cell in the medium. The signal-transduction pathway between the receptor and the cellular response involves phosphorylation events (see Chapter 4).

Smells

The noses of many animals are extremely sensitive to the presence of a vast range of chemicals. It is not always clear, however, exactly what it is about the particular chemical that stimulates a certain response. Molecules that on paper appear to be very closely related structural analogues often give very similar smell sensations, whereas chemicals that smell the same may have very differing structures. Despite this apparent paradox, receptors have been identified that are responsible for the detection of volatile compounds in the air. These receptors are found in the neuroepithethial cells of the nose and, like many others, are associated with a class of trimeric G proteins, G_{olf}, which have their influence on adenylyl cyclase.

Taste

A G protein analogous to transducin of the rods and cones of the eye is also involved in the transduction of the signal in the taste buds. The taste-specific G protein has been named gustducin.

Taste buds contain 40–60 specialised receptor cells organised into groups. Ligand binding, which results in the perception of taste, leads to a release of neurotransmitters from the taste cells caused by either depolarisation or hyperpolarisation of the cells. Four basic tastes are recognisable, each leading to a different cellular response. Salty tastes result from an increase in Na^+ movement through Na^+ channels, while sour tastes are the result of the blockage of K^+ or Na^+ channels by H^+. Sweet and bitter tastes, on the other hand, are mediated by the presence of G proteins. The receptor responsible for the sweet sensation leads to activation of a trimeric G protein, a rise in cAMP and cellular depolarisation. Some bitter compounds appear to result in G protein activation and a rise in intracellular Ca^{2+} ions, released from intracellular stores, the release probably being mediated by the inositol phosphate pathway.

As mentioned above, one of the G proteins involved is known as gustducin, G_g. Sequence data of this protein show that the rat form of the α subunit is approximately 80% identical to bovine transducin α subunit from the eye.

Auditory Perception

Many organisms are responsive to sound, often over a large range of frequencies. Reports that the activity of certain enzymes is correlated to the intensity of sound are fascinating. Studies in the rat have shown that the activity of guanylyl cyclase in the inner ear is inversely related to the volume of sound to which the animals are exposed, suggesting a new and exciting signal-transduction field which needs to be investigated.

However, the auditory system of vertebrates may be a good example of signal transduction without the use of second-messenger molecules. The mechanical force of the sound waves detected by hair cells opens the cell's ion channels via a direct mechanical mechanism. One of the important proteins involved here is myosin. However the system is under the influence of the Ca^{2+} ion concentration as Ca^{2+} chelators have a profound effect on the system and Ca^{2+} binding proteins such as calbindin-D_{28k} have been found to be associated with the auditory system of rats.

Summary

Perception of the environment is as vital to the whole organism as it is for individual cells. The response to many environmental factors involves the use of many signalling pathways and can be used to illustrate the coordination of signal transduction.

Many organisms respond to light, either tailoring their growth to optimise light capture as in plants, or to enable them to move around and avoid danger, as with animals. Plants monitor light by the use of phytochromes, UV-B, UV-A and blue light receptors.

One of the most well-characterised systems for detecting light is the mammalian eye. Here, photons of light cause the photoisomeriation of retinal, resulting in activation of the opsin protein, which in turn leads to activation of a trimeric G protein, G_t. The released $G_{t\alpha}$ causes the activation of a PDE, which catalyses the rapid breakdown of cGMP. The subsequent drop in cGMP in the cytoplasm leads to the closure of cation-specific channels in the plasma membrane and the formation of an electrochemical potential, propagated to the synaptic region of the cell. From here, electrical signals pass to the brain where the picture is deciphered.

Another environmental factor that needs to the monitored is the presence of chemicals. For prokaryotes this might include essential nutrients such as amino acids and sugars. Larger animals, such as mammals, can detect airborne chemicals as smells and soluble chemicals as taste. The sense of smell appears to use receptors, again linked to trimeric G proteins, having their effect through adenylyl cyclase. Taste also uses specialised receptor cells, but different types of taste appear to use different signal-transduction pathways.

The auditory system of vertebrates is a good example of direct signal transduction without the use of second messengers. However, the report that the guanylyl cyclase involved in the perception of sound in mammals is sensitive to the intensity of that sound opens up a new and fascinating area for investigation.

Further Reading

Photoreception

Ahmed, M. and Cashmore, A. R. (1993) HY4 gene of *A. thaliana* encodes a protein with characteristics of a blue-light photoreceptor. *Nature*, **366**, 162–6.

Benovic, J. L., Mayor, F., Somers, R. L., Caron, M. G. and Lefkowitz, R. J. (1986) Light dependent phosphorylation of rhodopsin by β-adrenergic receptor kinase. *Nature*, **321**, 869–72.

Bowler, C., Neuhaus, G., Yamagat, H. and Chua, N.-.H. (1994) Cyclic GMP and calcium mediate phytochrome phototransduction. *Cell*, **77**, 73–81.

Jelsema, C. L. and Axelrod, J. (1987) Stimulation of phospholipiase A_2 activity in bovine rod outer segmnets by the βγ subunits of transducin and its inhibiton by the α subunit. *Proceedings of the National Academy of Sciences USA*, **84**, 3623–7.

Lee, R. H., Brown, B. M. and Lolley, R. N. (1990) Protein kinase A phosphorylates retinal phosducin on serine 73 *in situ*. *Journal of Biological Chemistry*, **265**, 15860–6.

Lee, R. H., Ting, T .D., Lieberman, B. S., Tobias, D. E. Lolley, R. N. and Ho, Y.-K. (1992) Regulation of retinal cGMP cascade by phosducin in bovine rod photoreceptor cells. *Journal of Biological Chemistry*, **267**, 25104–12.

Neuhaus, G., Bowler, C., Kern, R. and Chua, N. H. (1993) Calcium/calmodulin-dependent and -independent phytochrome signal transduction pathways. *Cell*, **73**, 937–52.

Nöll, G. N., Billek, M., Pietruck, C. and Schmidt, K. F. (1994) Inhibition of nitric oxide synthase alters light responses and dark voltage of amphibian photoreceptors. *Neuropharmacology*, **33**, 1407–12.

Quail, P. H. (1994) Photosensory perception and signal transduction in plants. *Current Opinion in Genetic Development*, **4**, 652–61.

Ranganathan, R., Bacskai, B. J., Tsien, R. Y. and Zuker, C. S. (1994) Cytosolic calcium transients: spatial localisation and role in *Drosophila* photoreceptor cell function. *Neuron*, **13**, 837–48.

Stryer, L. (1986) Cyclic GMP cascade of vision. *Annual Review of Neuroscience*, **9**, 87–119.

Stryer, L. (1991) Visual excitation and recovery. *Journal of Biological Chemistry*, **266**, 10711–14.

Stryer, L. (1995) *Biochemistry* (4th edn). W. H. Freeman, New York. Chapter 13 has a particularly good summary of the signalling of the eye.

Detection of Chemicals

Amsler, C. D. and Matsumma, P. (1995) Chemotaxis signal transduction in *Escherichia coli* and *Salmonella typhimurium*. In *Two-component Signal Transduction*, Hoch, J. A. and Silhavy, T .J. (eds). American Society for Microbiology, Washington, 89–102.

Avenet, P. and Lindemann, B. J. (1989) Perspective of taste receptors. *Journal of Membrane Biology*, **112**, 1–8.

Kim, S.-H., Privé, G. G., Yeh, J., Scott, W. G. and Milburn, M. V. (1992) A model for transmembrane signalling in a bacterial chemotaxis model receptor. *Cold Spring Harbor Symposia of Quantitative Biology*, **57**, 17–24.

Kinnamon, S. C. (1988) Taste transducin: a diversity of mechanisms. *Trends in Neurosciences*, **11**, 491–6.

McLaughlin, S. K., McKinnon, P .J. and Margolskee, R.F. (1992) Gustducin is a taste-cell-specific G protein closely related to the transducins. *Nature*, **357**, 563–9.

Macnab, R. M. (1977) Bacterial flagella rotating in bundles: a study in helical geometry. *Proceedings of the National Academy of Sciences USA*, **74**, 221–5.

Maddock, J. R. and Shapiro, L. (1993) Polar localisation of the chemoreceptor complex in the *Escherichia coli* cell. *Science*, **259**, 1717–23.

Milburn, M. V., Privé, G. G., Milligan, D. L. *et al.* (1991) Three dimensional structures of the ligand-binding domain of the bacterial aspartate receptor with and without a ligand. *Science*, **254**, 1342–7.

Pakula, A. A. and Simon, M. I. (1992) Determination of the transmembrane protein structure by disulfide cross-linking: the *Escherichia coli* Tar receptor. *Proceedings of the National Academy of Sciences USA.*, **89**, 4144–8.

Roper, S. D. (1989) The cell biology of vertebrate taste receptors. *Annual Review of Neurosciences*, **12**, 329–53.

Stewart, R. C. and Dahlquist, F. W. (1987) Molecular components of bacterial chemotaxis. *Chemical Reviews*, **87**, 997–1025.

Auditory Perception

Friauf, E. (1994) Distribution of calcium-binding protein calbindin-D-28K in the auditory system of adult and developing rats. *Journal of Comparative Neurology*, **349**, 193–211.

Gillespie, P. G. (1995) Molecular machinery of auditory and vestibular transduction. *Current Opinion in Neurobiology*, **5**, 449–55.

Zubin, P., Defagot, C. and Parlanti, G. (1995) Guanylate cyclase activity in the inner ear and auditory nerve of the rat. *Journal of Biochemistry*, **118**, 418–21.

INDEX